Technological Change and the Future of Warfare

Technological Change and the Future of Warfare

MICHAEL O'HANLON

BROOKINGS INSTITUTION PRESS
Washington, D.C.

Copyright © 2000
THE BROOKINGS INSTITUTION
1775 Massachusetts Avenue, N.W., Washington, D.C. 20036
www.brookings.edu

Library of Congress Cataloging-in-Publication data

O'Hanlon, Michael E.
  Technological change and the future of warfare / Michael O'Hanlon.
      p.   cm.
  Includes bibliographical references and index.
  ISBN 0815764405 (cloth : alk. paper) - ISBN 0815764391 (paper :
alk. paper)
  1. United States-Military policy. 2. Military art and
science-Technological innovations. 3. War-Forecasting. I. Title.
  UA23 .O39 2000   99-050470
  355'.033573dc21   CIP

9 8 7 6 5 4 3 2 1

The paper used in this publication meets minimum requirements of the
American National Standard for Information Sciences—Permanence of Paper
for Printed Library Materials: ANSI Z39.48-1984.

Typeset in Myriad and Sabon

Composition by Oakland Street Publishing
Arlington, Virginia

Printed by R. R. Donnelley and Sons
Harrisonburg, Virginia

฿ THE BROOKINGS INSTITUTION

The Brookings Institution is an independent organization devoted to nonpartisan research, education, and publication in economics, government, foreign policy, and the social sciences generally. Its principal purposes are to aid in the development of sound public policies and to promote public understanding of issues of national importance.

The Institution was founded on December 8, 1927, to merge the activities of the Institute for Government Research, founded in 1916, the Institute of Economics, founded in 1922, and the Robert Brookings Graduate School of Economics and Government, founded in 1924.

The general administration of the Institution is the responsibility of a Board of Trustees charged with safeguarding the independence of the staff and fostering the most favorable conditions for scientific research and publication. The immediate direction of the policies, program, and staff is vested in the president, assisted by an advisory committee of the officers and staff.

In publishing a study, the Institution presents it as a competent treatment of a subject worthy of public consideration. The interpretations or conclusions in such publications are those of the author or authors and do not necessarily reflect the views of the other staff members, officers, or trustees of the Brookings Institution.

*To my parents and
the memory of my grandparents*

Foreword

This book examines in detail the currently popular hypothesis that it may be possible to achieve a revolution in military affairs (RMA) in the near future and that the U.S. military must radically reshape its budgetary priorities, warfighting structures, and weaponry in order to pursue it. Michael O'Hanlon describes the origins of the RMA debate and various schools of thought within it, while also providing historical perspective on how militaries try to innovate and what factors determine their success or failure.

O'Hanlon is skeptical about the RMA hypothesis in its grander forms and develops a number of technical, tactical, and strategic military arguments against it. He does find, however, that trends in technology justify focusing on how to fully exploit breakthroughs in electronics, computers, communications, and miniaturization—and placing less emphasis on modernizing large weaponry such as tanks, ships, and planes.

The RMA movement offers other recommendations for U.S. defense policy apart from the budgetary and organizational. O'Hanlon challenges several of the most popular of these as well. He finds that trends in technology suggest overseas U.S. military bases will remain important well into the future. Also, he is optimistic that trends in technology need not increase the gap in military capabilities between the United States and its major allies. Finally, he is highly pessimistic that it will be possible to verifiably eliminate nuclear and biological weapons from the earth, and thus throws some cold water on the nuclear abolitionist movement that has recently gained strength in the United States.

O'Hanlon is grateful to a large number of individuals from the Washington think tank world, the Pentagon, the national defense laboratories, and academia. He thanks Ronald Atlas, Michael Berger, Bruce Blair, Jerome Bracken, Robin Buckelew, Ivo Daalder, Thomas Davis, Patrick Eicker, George Fenton, Frank Finelli, Kenneth Flamm, Richard Garwin, Thomas Garwin, Bates Gill, Walter Givhan, Philip Gordon, Michael Green, Robert Haffa, Houston Terry Hawkins, John Hillen, Frank Hoffman, Susan Jackson, Mim John, Mark Johnson, John Kennedy, Charles Krulak, Richard Kugler, Valerie Lopez, Ken McKenna, Thomas McNaugher, Thomas Meyer, Randy Michelsen, Mike Mochizuki, Steve Newfield, Janne Nolan, David Ochmanek, Lane Pierrot, Joseph Polito, Ken Pollack, Barry Posen, Robert Rinne, Daniel Rondeau, David Russell, Steve Sargeant, Woody Shortt, John Steinbruner, Mark Strauss, Mohammed Sulaiman, Gordon Sullivan, John Taylor, David Whelan, Jerre Wilson, John Wissler, Ainslie Young, and others at the national laboratories, the Pentagon, and elsewhere. He is especially grateful to Brian Finlay and Julien Hartley, who provided research assistance; to Richard Betts, Stephen Biddle, Andrew Marshall, David Mosher, and Frank von Hippel, who read drafts of the manuscript and commented in detail; and to his wife, Cathryn Garland. The work was performed under the direction of Richard Haass, who also provided very helpful comments. Of course, none of these people is responsible for the opinions, or any mistakes, found in these pages.

Brookings is grateful to the Ford Foundation, the U.S. Institute of Peace, and other donors for their financial support of this project.

MICHAEL H. ARMACOST
President

Washington, D.C.
January 2000

Contents

CHAPTER ONE

Introduction 1

CHAPTER TWO

The So-Called Revolution in Military Affairs 7

CHAPTER THREE

Sensors, Computers, and Communications 32

CHAPTER FOUR

Vehicles, Ships, Aircraft, and Weaponry 68

CHAPTER FIVE

Toward a Verdict on the RMA Hypothesis 106

CHAPTER SIX

The RMA Hypothesis and
U.S. Security Policy 143

CHAPTER SEVEN

A Defense Modernization Strategy 168

CHAPTER EIGHT
Conclusion   192

Index   199

Tables
4-1. Visibility of Selected Objects to Radar   73
4-2. Current and Planned U.S. Unmanned Aerial Vehicles   75
4-3. Maximum Speed of Destroyers over Time   80
4-4. Demonstration and Validation Budget for Currently   92
     Funded Joint Nonlethal Weapons Programs, FY 2000
7-1. Major Pentagon Acquisitions Programs, August 1999   187
Box
5-1. Summary of Key Technological Assumptions   109
Figures
3-1. Reflectivity of U.S. and Soviet Camouflage   35
     and Healthy Green Vegetation
3-2. Evolution of Computing Power   56
3-3. Projected Advances in Key Military   65
     Technologies, 2000–20
4-1. Aircraft Speed since World War II   70
4-2. The Benefits of Stealth   72
4-3. Theater Missile Defense Architecture   98
5-1. Historical Army Dispersion Patterns   122
7-1. International Distribution of Military   169
     Spending, Selected Countries, 1998
7-2. U.S. Defense Procurement Spending, 1945–2003   170
7-3. U.S. National Defense Spending, 1962–2004   171
7-4. U.S. Defense Research, Development, Test,   183
     and Evaluation Spending, 1945–2005
7-5. U.S. Department of Defense Basic Research   184
     Budget, 1972–2000
7-6. U.S. Department of Defense Applied Research   185
     Budget, 1972–2000
7-7. Quality of Fighter Aircraft, 1995, 2005,   190
     and 2015

Technological
Change and the
Future of Warfare

CHAPTER ONE

Introduction

In the second half of the twentieth century, high technology became the defining characteristic of the American way of war. It is certain to remain central to U.S. defense policy in the twenty-first century as well. American military personnel are also outstanding. But it is their juxtaposition with the world's best defense technology that has created the finest military force in history.

High technology has not always been such a central ingredient in U.S. defense strategy. In the world wars, it was less the caliber of U.S. military technology than its sheer quantity that provided the raw material for victory. To be sure, that quantity was itself a function of American technological supremacy in the form of mass-production industry. But after World War II, U.S. defense strategists placed primary emphasis on ensuring that American military equipment would be of superior quality.

This transformation occurred gradually. In Vietnam, U.S. forces benefited from a wide array of new defense systems, ranging from helicopters to satellites to high-performance jets to laser-guided bombs. But the emphasis remained on massing weaponry and firepower. Not until the invasion of Panama in 1989 and Operation Desert Storm in 1991 was the transition to a high-tech American military truly complete.

In a broader sense, however, no such transition can ever be considered complete. Technology is constantly advancing—particularly in a world that is systematically organized to conduct scientific and engineering research on a large scale. The armed forces of a country, such as the United States, that depends heavily on technology must innovate constantly in order to stay ahead. In addition, as has been underscored by the tragic U.S. experi-

1

ence in Somalia in 1993, inconclusive aerial attacks against Iraq in 1998, the human cost for Kosovar Albanians of the long—if ultimately successful—NATO air war against Serbia in 1999, and the vulnerabilities of modern societies to missile and terrorist attack, our technological edge has its limitations even today.

With these historical trends and future policy challenges in mind, many American defense analysts have recently posited that a revolution in military affairs (RMA) is either under way or within reach and that the United States needs to be aggressive about pursuing it. Likening this time in geopolitics and military history to the 1930s or early 1940s, they see great potential for radically new military hardware and operational concepts—as well as great dangers if another country transforms its forces and realizes the revolution's promise before we do.

As discussed further in chapter 2, there are many different visions of what a contemporary revolution in military affairs should entail and how the United States should try to usher it in. It would be wrong to imply that all those who promote the RMA concept have the same idea of what it means today. Nonetheless, a common definition of the purported RMA is emerging, based in large part on how the Pentagon—now an official convert to the RMA concept—has chosen to define the term. Former chairman of the Joint Chiefs of Staff John Shalikashvili's *Joint Vision 2010* report and the subsequent 1997 Quadrennial Defense Review conducted under Defense Secretary William Cohen have figured prominently in the effort to define a contemporary RMA. Most notable are the goals these documents set out for U.S. military forces a decade into the twenty-first century: that they possess "dominant battlefield knowledge," "full-dimensional protection," "dominant maneuver," and "precision strike" ability from long distances. But what do these terms really mean? My understanding of the *Joint Vision 2010* RMA hypothesis, and those of similar schools of thought, is that they accept the following specific technological premises:

—First, improvements in computers and electronics will make possible major advances in weapons and warfare—most notably in areas such as information processing and information networks but also in communications, robotics, advanced munitions, and other technologies.

—Second, sensors will become radically more capable, in effect making the battlefield "transparent."

—Third, land vehicles, ships, rockets, and aircraft will become drastically lighter, more fuel efficient, faster, and more stealthy, making combat forces far more rapidly deployable and lethal once deployed.

—Fourth, new types of weaponry—such as space weapons, directed energy beams, and advanced biological agents—will be developed and widely deployed.

They also posit two sweeping conclusions:

—First, if properly exploited and integrated into military organizations, tactics, and concepts of operations, these technical trends can soon add up to a revolution in military affairs that will constitute the greatest advances in warfare since the advent of blitzkrieg and aircraft carriers in the 1930s and nuclear weapons in the 1940s.

—Second, U.S. adversaries, even if considerably less technologically sophisticated and wealthy than the United States, will also benefit from this contemporary RMA. Notably, they will acquire and learn to make good use of advanced precision missiles, satellites, antisatellite weapons, advanced mines, weapons of mass destruction, and computer viruses—and thus be able to challenge U.S. operations much more than Iraq did in Operation Desert Storm or Serbia did in Operation Allied Force. They are particularly likely to exploit the U.S. military's dependence on large bases, ships, and other vulnerable assets when projecting power overseas, as well as Americans' aversion to suffering casualties. As a result, the United States needs to seek radically new military concepts to overcome these challenges to its military supremacy and indeed its basic security.

A major contention of this book is that, while the first technological premise of the RMA hypothesis is essentially correct, the second and third are incorrect or at least badly overstated by their proponents. The fourth premise is less easily evaluated at this point, but there are ample grounds for skepticism.

Most mechanical technologies that are central to military systems have not been changing as fast as electronic and computational technologies; nor are they likely to do so in the first couple decades of the twenty-first century. In addition, although some types of sensors will improve substantially through a variety of engineering advances, their potential will generally remain limited by basic laws of physics and by an enemy's ability to take advantage of countermeasures as well as the cover provided by natural and civilian backgrounds.

As for the broad conclusions, these technological realities and trends suggest that any contemporary revolution in military affairs, should one prove possible, would have to be driven by developments in a fairly narrow subset of major defense technologies. One cannot rule out an RMA on these grounds alone. But they do suggest that the likelihood of an RMA is less, and the probable speed and scope of any near-term revolution less sweeping, than proponents commonly admit—particularly for the types of conflicts that seem most likely to involve the United States in coming decades.

The second conclusion is more persuasive, if only because the history of warfare is about enemies trying to bypass each other's strengths and exploit each other's vulnerabilities. It is only natural that potential and actual U.S. adversaries will try to make widespread use of new technology themselves, and that—given their (generally inferior) means—they will do so in asymmetric ways, attempting to exploit specific U.S. vulnerabilities rather than defeat the American armed forces head-on in combat. However, enemies of the United States will not be able to achieve these asymmetric capabilities as easily as some RMA believers suggest. Nor will the best U.S. response to such enemy capabilities always require radically new high technology, as RMA believers tend to argue. Nor will radically new technologies always be achievable, even if the United States tries to develop them.

This book is not only an assessment of the contemporary RMA hypothesis. It is also intended as a more specific analysis of how the American military should acquire defense technology in the years ahead. Whether one believes a revolution is under way—and hence whether one believes the basic structure, equipment, and operational concepts of the U.S. armed forces must be radically overhauled soon—it is true that future decisions about weaponry and warfighting are of major importance for American security. The country must determine how much to spend on defense, how many resources to put into weapons modernization, and how to apportion those modernization resources between technological research and development (R&D) on the one hand and weapons procurement on the other.

When the issue is put in these terms—which are more pragmatic and less heady than RMA proponents generally prefer—it becomes very clear how important it is to determine which of the four RMA technological premises are correct. If, as I contend, the technologies undergirding most types of military vehicles and major weapons platforms are not advancing at revolutionary rates, it does not make sense to rapidly replace or transform these vehicles and platforms wholesale. Defense modernization can instead focus largely on R&D, war gaming and experimentation, and a targeted and relatively economical procurement strategy. Moreover, this modernization effort can probably be afforded, even at current defense budget levels, without reducing U.S. military operations in Korea, the Persian Gulf region, Bosnia, Kosovo, and the vicinity of the Taiwan Strait.

If there is indeed to be a contemporary revolution in military affairs, a great number of new technologies, warfighting concepts, and organizational innovations are still required to make it possible. A radical transformation may ultimately be appropriate, but it is too soon to know this—and certainly too soon to carry it out by radically reshaping the military and reequipping it essentially from scratch. The early years of the twenty-first century are thus more akin to the 1920s than the 1930s, to echo Pentagon Director of Net Assessment Andrew Marshall and Commandant of the Marine Corps James Jones.[1] In fact, the first decade or two of the twenty-first century might be most akin to the last five of the twentieth century, when the United States systematically pursued defense innovation and a gradual, yet very real, military transformation without adopting a strategy for doing so urgently.

This study's primary focus is the sphere of traditional warfighting. Subjects such as homeland defense and strategic information warfare are touched upon as well; however, this book emphasizes the arena of combat in which countries attempt to defeat each other's main military forces— and most notably each other's conventional military forces. That said, it is not only about large-scale combat that features heavy weapons, but also about urban and forest warfare, humanitarian intervention, peace operations, and other missions that may play less directly to U.S. military strengths.

This book begins with an overview of the RMA hypothesis and a brief historical sketch of past revolutions in military affairs. The rest of the book falls into three main substantive sections.

Chapters 3 and 4 survey trends in key areas of military technology. Not only computers and communications equipment, but also various types of sensors as well as military vehicles and weapons are considered. RMA literature to date has generally failed to provide such a systematic assessment of where defense technology is headed; instead, it has based its reasoning largely on anecdotes or selective use of statistics on computer and modern electronics systems advancements. This section provides the basis for evaluating the validity of the four RMA technological premises.

Some will object to my technology-oriented methodology, claiming that revolutions in military affairs are less accidents of invention than the pur-

---

1. See Jay Winik, "Secret Weapon," *Washingtonian*, vol. 34 (April 1999), p. 48; J. L. Jones, Marine Corps Operations Deputy to the Joint Chiefs of Staff, "Memorandum for the Director of the Joint Staff: Evaluation of Service RMA Activities," PLN MCODM 001-97, January 13, 1997.

poseful creations of military establishments. Although they are largely right on the latter point, technology has been an essential ingredient in most RMAs. For example, at least one major development in weaponry contributed centrally to each of the ten major military revolutions since 1300, as identified by Andrew Krepinevich.[2] Moreover, most who argue that an RMA is afoot or within reach today begin their case with reference to recent radical leaps in the power of computers and electronics, often claiming to be able to extrapolate those trends to other areas of defense technology.

In chapter 5 (the second section), the technological prognostications of chapters 3 and 4 are integrated to examine the implications of future military technology for various types of warfighting scenarios. This chapter also puts forth a broad verdict on the RMA hypothesis, accepting parts of it but disputing others, and arguing against the most sweeping policy prescriptions of its proponents.

Chapters 6 and 7 (the third and final section) consider the policy implications of evolving defense technology. They consider security issues such as U.S. requirements for overseas military bases in the early twenty-first century and the future of multinational military operations. They also offer prescriptions for Pentagon budgeting. I argue that near-term defense investment can be discriminating and patient, emphasizing systems that make particularly high use of electronics, computers, and advanced communications systems rather than major weapons platforms and vehicles. Scientific progress is most rapid in the spheres of electronics and computers. RMA proponents are right about the remarkable trends in these areas, even if they are often wrong to suggest that other areas of defense technology are advancing almost as fast. Therefore, with this investment strategy, the Pentagon can achieve considerable improvements in military capability at modest cost, allowing the nation to keep defense spending at current levels without having to sacrifice current military operations or skimp on research aimed at the future.

The Pentagon should sustain vigorous R&D and experimentation activities. Someday those activities, and the technologies and new warfighting concepts they produce, may convincingly argue for a major overhaul of the U.S. military. We cannot yet know if that will be the case. But we can be confident that there is not a compelling case for immediate and wholesale transformation at the turn of the century.

2. Andrew Krepinevich Jr., "Cavalry to Computer: The Pattern of Military Revolutions," *National Interest*, no. 37 (Fall 1994), pp. 30–42.

CHAPTER TWO

The So-Called Revolution in Military Affairs

In light of the spectacular performance of American high-technology weapons in the 1991 Persian Gulf War, as well as the phenomenal pace of innovation in the modern computer industry, many defense analysts have posited that we are on the threshold of a revolution in military affairs (RMA). The RMA thesis holds that further advances in precision munitions, real-time data dissemination, and other modern technologies can help transform the nature of future war and with it the size and structure of the U.S. military. RMA proponents believe that military technology, and the resultant potential for radically new types of warfighting tactics and strategies, is advancing at a rate unrivaled since the 1930s and 1940s, when blitzkrieg, aircraft carriers, large-scale amphibious and airborne assault, ballistic missiles, strategic bombing, and nuclear weapons were developed.

The RMA movement is not motivated only by the allure of modern technology; other factors are at work as well. Some believe that only a high-technology standoff-warfare force can make the U.S. military usable in a domestic political context, given Americans' aversion to suffering casualties.[1] This philosophy guided NATO's 1999 war over Kosovo, in which the United States lost no troops to hostile action and only two people in the entire operation—largely as a result of the decisions to eschew an early ground invasion and to fly combat missions from high altitudes. Other RMA proponents argue that established powers are vulnerable to settling

---

1. Edward N. Luttwak, "A Post-Heroic Military Policy," *Foreign Affairs*, vol. 75 (July–August 1996), pp. 33–44.

into strategic complacency, gradually losing ground relative to other countries as a result, and that the United States, therefore, needs to find a new
catalyst to innovation given the end of the cold war.[2] If an RMA debate can
provide that catalyst, there is much to be said for it. Another group of RMA
supporters believes that the world is considerably less dangerous today than
it has been—or than it probably will be in the near future. Defining the turn
of the century as a moment of "strategic pause," they advocate using this
respite from serious international rivalry to remake the armed forces—in
part, because of the promise of modern technology, and, in part, because of
the absence of an immediate threat. Nevertheless, at its core the contemporary revolution in military affairs movement is motivated by technology and
its potential implications for warfare.

RMA proponents tend to be somewhat anti-Clausewitzian. Unlike the
nineteenth-century German strategist Carl von Clausewitz, who coined the
famous phrase "fog of war" to describe the unpredictability and confusion
of battle, they believe that future militaries should attempt to achieve information dominance—and that a winning force will probably succeed in
establishing it. They believe that future militaries will be able to depend on
highly complex and integrated communications systems that enable them
to fight in cohesive and complex ways. That basic concept tends to run
counter to Clausewitzian axioms that, in war, things break, seemingly easy
activities become slow and difficult, initial battle plans must usually be discarded, and human character becomes as important as intelligence, technology, or strategy.[3]

Being anti-Clausewitzian may not be all bad. After all, the German armies
that executed the brilliant blitzkrieg operations of World War II had themselves discarded his advice, insisting on speed and cohesion in their attack
plans. The U.S. military took a similar attitude in the 1991 war against Iraq.
Moreover, RMA proponents have their own famous military theorists to
invoke for inspiration and validation. Most notable among these are the
ancient Chinese strategist Sun Tzu and the twentieth-century British mili-

2. See, for example, Aaron L. Friedberg, *The Weary Titan: Britain and the Experience of
Relative Decline, 1895–1905* (Princeton University Press, 1988), pp. 299–303.

3. See Admiral William A. Owens, "Foreword," in Stuart J. D. Schwartzstein, *The Information Revolution and National Security* (Washington, D.C.: Center for Strategic and International Studies, 1996), pp. xi–xii; Michael J. Mazarr, with Jeffrey Shaffer and Benjamin
Ederington, *The Military Technical Revolution* (Washington, D.C.: Center for Strategic and
International Studies, 1993), p. 23; on Clausewitz, see the translation by Michael Howard
and Peter Paret of Carl von Clausewitz, *On War* (Princeton University Press, 1989),
pp. 119–21.

tary scholar B. H. Liddell Hart. Sun Tzu wrote of the desirability of achieving victories with minimal bloodshed and short battles, avoiding enemy strengths with agility and deception, and winning through savvy and skill rather than brute force.[4] Liddell Hart advocated an indirect approach to warfare that emphasized maneuver, deception, and, above all, the avoidance of pitched battle against prepared defenses.[5]

Although RMA advocates in the United States tend to see the Persian Gulf War as the harbinger of a contemporary revolution in military affairs, it was hardly the first impressive display of modern military technology. For example, in just eighteen days of fighting in the October 1973 Arab-Israeli War, advanced weapons proved so lethal that the participating militaries lost roughly one-third of their initial major weapons systems.[6] Laser-guided bombs had made their debut somewhat sooner—in the Vietnam War.

These and other developments made a strong impression on defense experts like the Soviet Union's chief of the general staff in the early 1980s, Marshal General Nikolai Ogarkov. Indeed, already in the 1970s, some Russians were coining the expression military-technical revolution to capture ongoing progress in areas such as jet engine technology, electronics, advanced sensors, and precision munitions.[7] Many of these technologies and ideas, as well as the development of stealth shapes and coatings for aircraft, influenced the U.S. military's follow-on forces attack (FOFA) concept for fighting against the Warsaw Pact in Europe in the mid-1980s.[8]

It was the high-tech success of the Persian Gulf War, however, that brought the idea of a contemporary revolution in military affairs to the forefront of U.S. defense debates. Conclusive data on weapons performances during that conflict remain elusive, largely because many of the targets were never

4. See Samuel B. Griffith's translation of Sun Tzu, *The Art of War* (Oxford University Press, 1963), pp. 63–110.

5. B. H. Liddell Hart, *Strategy*, 2d ed. (New York: Meridian Books, 1991), pp. 207–37, 319–52.

6. Kenneth L. Adelman and Norman R. Augustine, *The Defense Revolution: Strategy for the Brave New World* (San Francisco, Calif.: Institute for Contemporary Studies, 1990), p. 54.

7. Nikolai A. Lomov, ed., *Scientific-Technical Progress and the Revolution in Military Affairs (A Soviet View)*, translated and published under the auspices of the U.S. Air Force (Government Printing Office, 1974), pp. 74–98.

8. Office of Technology Assessment, *New Technology for NATO: Implementing Follow-on Forces Attack* (Government Printing Office, 1987), pp. 3–52; Eliot A. Cohen, "A Revolution in Warfare," *Foreign Affairs*, vol. 75 (March–April 1996), p. 39; François L. Heisbourg, "Conventional Defense: Europe's Constraints and Opportunities," in Andrew J. Pierre, ed., *The Conventional Defense of Europe: New Technologies and New Strategies* (New York: Council on Foreign Relations, 1986), pp. 79–85.

inspected after the war (being inside Iraq or otherwise unavailable).[9] However, a number of notable accomplishments can be documented with reasonably good confidence.

In Operation Desert Storm, U.S. airpower destroyed from 2,000 to 3,000 pieces of heavy armored equipment, despite flying at altitudes of more than two miles to avoid groundfire.[10] Assuming that precision munitions did most of the damage, as appears likely given the hardness of the targets and probable ineffectiveness of inaccurate ordnance against them, that translates into an estimated kill probability per advanced munition of roughly 20 to 30 percent—several times what had been viewed as exceptional in previous warfare.[11] Tanks and other vehicles could be targeted even though they were partially hidden in various prepared positions and even though the Iraqis controlled the surrounding ground.

Western armored forces in Desert Storm performed extremely well, too. For example, the combination of thermal imaging equipment, laser rangefinders, stabilized gun barrels, and depleted uranium shells allowed U.S. tanks to strike Iraqi targets from two to three kilometers away with as high as 85 percent kill probability. Moreover, they often did so while on the move and operating in sandstorms. The resulting exchange ratio of perhaps 50:1 in favor of U.S. equipment and personnel was at least five times as good as Israeli armies had typically obtained against Arab foes in earlier wars.[12]

RMA proponents disagree as to whether the Persian Gulf War was a harbinger of long-range high-precision warfare or the first signs of its culmination.[13] However, most of those who promote the concept in today's

9. See General Accounting Office, *Operation Desert Storm: Evaluation of the Air Campaign*, GAO/NSIAD-97-134 (June 1997), pp. 110–18, 146–61.

10. General Accounting Office, *Operation Desert Storm*, pp. 147–48; Eliot A. Cohen, *Gulf War Air Power Survey*, vol. 2 (Government Printing Office, 1993), part 2, pp. 170, 214–19; Robert A. Pape, *Bombing to Win: Air Power and Coercion in War* (Cornell University Press, 1996), p. 251.

11. Thomas A. Keaney and Eliot A. Cohen, *Gulf War Air Power Survey Summary Report* (Government Printing Office, 1993), p. 103; General Accounting Office, *Operation Desert Storm*, p. 218; Barry R. Posen, "Measuring the European Conventional Balance: Coping with Complexity in Threat Assessment," in Steven E. Miller, ed., *Conventional Forces and American Defense Policy* (Princeton University Press, 1986), p. 104.

12. Stephen Biddle, "Victory Misunderstood: What the Gulf War Tells Us about the Future of Conflict," *International Security*, vol. 21 (Fall 1996), pp. 156, 161; Posen, "Measuring the European Conventional Balance," p. 113; Trevor N. Dupuy, *Attrition: Forecasting Battle Casualties and Equipment Losses in Modern War* (Fairfax, Va.: Hero Books, 1990), p. 139.

13. See Eliot A. Cohen, Michael J. Eisenstadt, and Andrew J. Bacevich, *Knives, Tanks, and Missiles: Israel's Security Revolution* (Washington Institute for Near East Policy,

American defense debate believe that the so-called revolution in military affairs is far from over. They also generally think that the country must consider fairly radical measures in order to realize that revolution.

The RMA Hypothesis

What do people mean, specifically, when they say that a revolution in military affairs is either under way or within reach? In fact, while definitions do vary from person to person, there is one understanding of the term that is fairly widespread. It is reflected in the Pentagon's current rhetoric and official documents as well as a host of writings of independent scholars and strategists. This definition of a contemporary RMA is rather ambitious, encompassing all four of the technological premises and both of the broad conclusions laid out in chapter 1. It is also rather unrealistic. As such, it is analytically useful to break down this broad definition of an RMA into constituent elements or schools of thought, some of which are more realistic and better guides to policy than others.

RMA Schools of Thought

At a conceptual level, there are perhaps four main schools of RMA thought. The first three are progressively more bullish in their RMA enthusiasm, the last school is of a different type.

—The *system of systems* school focuses on the potential of rapidly improving computers, communications, and networking to make existing weapons systems function in a much more integrated fashion. It accepts the first technological premise put forth in chapter 1, but not the others.

—The *dominant battlespace knowledge* school accepts the premises of the system of systems school, but also assumes radical improvements in sensors that will make future battlefield data much better and more complete. In other words, it endorses the first and second technological premises as described in chapter 1.

1998), p. 10; Norman C. Davis, "An Information-Based Revolution in Military Affairs," in John Arquilla and David Ronfeldt, eds., *In Athena's Camp: Preparing for Conflict in the Information Age* (Santa Monica, Calif.: RAND, 1997), p. 85; Jeffrey R. Cooper, "Another View of the Revolution in Military Affairs," in Arquilla and Ronfeldt, *In Athena's Camp*, p. 128.

—The *global reach, global power* school accepts the hypotheses of the system of systems and dominant battlespace knowledge schools, but also envisions the development of far more lethal, agile, and deployable weapons. Members of this RMA school of thought accept most or all of the technological premises outlined in chapter 1.

—Finally, the less confident *vulnerability* school posits that adversaries may benefit at least as much as the United States from technologies like advanced sea mines, submarines, cruise missiles, ballistic missiles, satellite imaging, computer viruses, radio-frequency weapons, antisatellite weapons, and weapons of mass destruction. In other words, this school emphasizes the second broad conclusion of the RMA movement as proposed in chapter 1.

SYSTEM OF SYSTEMS. Virtually all contemporary RMA visions emphasize the concept of a system of systems: that future warfare will be dominated less by individual weapons platforms and munitions than by real-time data processing and networking that tie U.S. forces together synergistically. Proponents point to the fact that computers have been getting much faster for years. Supercomputer computational power has been increasing by a factor of ten every five years.[14] Personal computers have improved almost as quickly, roughly doubling in speed every two years since IBM's personal computer was introduced in 1981.[15] Although the computer's benefits for the economy were unclear for the 1980s and the early 1990s, recent economic evidence suggests that information technology may be largely responsible for the prolonged U.S. economic expansion of the mid- to late 1990s. If this effect is real and sustainable, perhaps computers will soon be just as beneficial for military operations.[16]

Trends in computing power, speed, cost, and size have made it possible to put computers on ballistic missiles, fighter jets, and phased-array radars in the last few decades. Further advancements now make it possible to put computing capability on all significant platforms and to network the systems together. This will allow such systems to gather information from many sources, process it in real time, and rapidly exchange data on the bat-

14. Kenneth Flamm, "Controlling the Uncontrollable," *Brookings Review*, vol. 14 (Winter 1996), pp. 22–25.

15. Martin Libicki, "Technology and Warfare," in Patrick M. Cronin, *2015: Power and Progress* (National Defense University Press, 1996), p. 120.

16. On the reasons for skepticism, see Stephen Biddle, "The Past as Prologue: Assessing Theories of Future Warfare," *Security Studies*, vol. 8 (Autumn 1998), pp. 34–44; on the recent good economic news, see Steve Lohr, "Computer Age Gains Respect of Economists," *New York Times*, April 14, 1999, p. A1.

tlefield.[17] To put it differently, radical progress is under way in C4—or command, control, communications, and computers—technologies, and the U.S. military should be able to derive great benefits from that progress. The system of sytems phrase was popularized by Admiral William Owens, who served as vice chairman of the joint chiefs of staff in the mid-1990s.[18]

DOMINANT BATTLESPACE KNOWLEDGE. Many of those who accept the system of systems concept expect even more from future military technology. Convinced that radical improvements are under way not only in computers but also in sensors that gather information, they have invoked the phrase dominant battlespace knowledge (DBK) to describe a future combat environment in which the United States would be able to promptly find and continuously track virtually all important enemy assets within a combat zone often specified as being 200 nautical miles square. (That is roughly the size of key battlefield areas in a place such as Kuwait or the Korean peninsula.)

As its name suggests, the DBK school is much more bullish and ambitious than the system of systems school. It not only presupposes the rapid processing and exchange of information on the battlefield, but also the availability of much better information to process and exchange.[19] In other words, it expects breakthroughs not only in C4 technologies, organizations, and capabilities, but also huge strides in intelligence, surveillance, and reconnaissance (ISR), making for a complete C4-ISR revolution in military affairs. As one prominent proponent, former chief of staff of the air force Ronald Fogleman, put it before Congress in 1997: "In the first quarter of the 21st century you will be able to find, fix or track, and target—in near real-time—anything of consequence that moves upon or is located on the face of the Earth."[20]

Clearly those who subscribe to the more limited system of systems concept understand that sensors will continue to improve. For example, the

17. Martin C. Libicki, "DBK and Its Consequences," in Stuart E. Johnson and Martin C. Libicki, eds., *Dominant Battlespace Knowledge* (National Defense University Press, 1996), pp. 23–49.

18. See, for example, Owens's coauthored article: Joseph S. Nye Jr. and William A. Owens, "America's Information Edge," *Foreign Affairs*, vol. 75 (March–April 1996), pp. 23–25; see also Jack Dorsey, "Now's Time to Take Risks for New Navy, Admiral Says," *Norfolk Virginian-Pilot*, October 26, 1997, p. 1; "Owens: Get Smart Weapons," *Navy News and Undersea Technology*, October 3, 1994, p. 1.

19. See Johnson and Libicki, *Dominant Battlespace Knowledge*.

20. See Statement of General Ronald R. Fogleman, Chief of Staff, U.S. Air Force, before the House National Security Committee, 105 Cong. 1 sess., May 22, 1997.

miniaturization of electronics, on-board information-processing capabilities, GPS receivers, and secure high-data-rate radios now make possible devices like unmanned aerial vehicles (UAVs). In addition, improvements or innovations in sensors will probably take place in areas such as multi-spectral imaging and foliage-penetrating radar, which will be important in certain specific contexts. Proponents of the system of systems concept do not, however, anticipate that sensors will improve so drastically as to make the battlefield transparent.

GLOBAL REACH, GLOBAL POWER. Certain schools of thought place a heavy premium on new types of weaponry to deliver ordnance extremely fast and in new ways. Proponents of this type of vision contemplate being able to base forces in the United States but deploy them rapidly and decisively overseas within hours or at most a few days; they also see the United States being able to avoid dependence on large fixed bases in combat theaters.

The U.S. Air Force first coined the phrase global reach, global power, and used it to argue for more resources for certain types of air force programs.[21] Given its dominant role in winning the Persian Gulf War—not to mention Operation Allied Force against Serbia in 1999—this is not surprising. Some additional attributes of air force–oriented force postures are that they promise wars with few U.S. casualties and a rapid U.S. military response to crises or conflicts virtually anywhere on earth.

Although these air force visions vary, they generally emphasize the firepower and rapid-response capabilities of systems such as stealthier air-to-air fighters, B-2 bombers, advanced reconnaissance capabilities such as UAVs, and "brilliant" munitions like the sensor-fuzed weapon (SFW) with autonomous terminal homing capabilities that do not require human operators in their final approach to a target.[22]

21. Sheila E. Widnall, "Report of the Secretary of the Air Force," in William S. Cohen, *Annual Report to the President and the Congress* (Department of Defense, 1997), p. 270.

22. See, for example, Christopher Bowie and others, *The New Calculus: Analyzing Airpower's Changing Role in Joint Theater Campaigns* (Santa Monica, Calif.: RAND, 1993); David A. Ochmanek and others, *To Find, and Not to Yield: How Advances in Information and Firepower Can Transform Theater Warfare* (Santa Monica, Calif.: RAND, 1998); Charles M. Perry, Robert L. Pfaltzgraff Jr., and Joseph C. Conway, *Long-Range Bombers and the Role of Airpower in the New Century* (Cambridge, Mass.: Institute for Foreign Policy Analysis, 1995); Zalmay Khalilzad and David Ochmanek, "Rethinking U.S. Defense Planning," and Benjamin S. Lambeth, "The Technology Revolution in Air Warfare," *Survival*, vol. 39 (Spring 1997), pp. 43–64 and 65–83, respectively; Daniel Goure and Stephen A. Cambone, "The Coming of Age of Air and Space Power," in Daniel Goure and Christopher M. Szara, eds., *Air and Space Power in the New Millennium* (Washington, D.C.: Center for Strategic and International Studies, 1997), pp. 1–47.

These air force–related visions sometimes include specific force structure proposals that would require cuts in the other services and entrust the air force with more than the 30 percent of total Pentagon resources it has typically received over the last three decades.[23] As such, they make the air force few friends within army, navy, and Marine Corps ranks. Be that as it may, air force proponents offer specific suggestions that can be scrutinized and evaluated. The RMA debate, as well as the general U.S. defense debate, needs such unencumbered proposals. The alternative is to give each of the military services their standard share of the defense budget—in essence making defense strategy in the comptroller's office.[24]

The concept of global reach and global power goes well beyond the air force, however. For example, some envision that ground combat units will be organized in radically different ways, permitting them to deploy very rapidly with only modest amounts of equipment and supplies. They might function in very small mobile teams that conduct tactical reconnaissance and call in precise strikes from distant ships or aircraft as they locate enemy assets difficult to identify from air or space. According to a 1996 Defense Science Board task force: "There is a good chance that we can achieve dramatic increases in the effectiveness of rapidly deployable forces if redesigning the ground forces around the enhanced combat cell [light, agile units with 10 to 20 personnel each] proves to be robust in many environments. There is some chance all this will amount to a true revolution in military affairs by 'eliminating the reliance of our forces on the logistics head as Blitzkrieg freed the offense after World War I from its then decades old reliance on the railhead.'"[25]

The U.S. Marine Corps espouses a related concept. The corps wishes to make future units smaller and to base much of their logistics support on ships or perhaps on mobile offshore bases with enormous carrying capacity, airstrips, and resilience to attack. Those capabilities, combined with longer-range airpower such as the MV-22 Osprey tilt-rotor aircraft, would supercede the traditional marine notion of storming the beach, purportedly allowing the Marines to keep many weapons and logistics assets at sea while

23. Office of the Under Secretary of Defense (Comptroller), *National Defense Budget Estimates for FY 2000* (Department of Defense, March 1999), pp. 138–41.

24. M. Thomas Davis, *Managing Defense after the Cold War* (Washington, D.C.: Center for Strategic and Budgetary Assessments, June 1997), p. iv; John Hillen, "Defense's Death Spiral," *Foreign Affairs*, vol. 78 (July–August 1999), pp. 2–7.

25. See Defense Science Board 1996 Summer Study Task Force, *Tactics and Technology for 21st Century Military Superiority*, vol. 1 (Department of Defense, 1996), p. S-4.

sending maneuver and scout forces deep into enemy territory directly from their ships.[26] Recently, the army has gotten into the act as well, with Chief of Staff General Eric Shinseki promoting acquisition of armored vehicles only one-third as heavy as today's that would erase the distinction between light and heavy forces, eliminate tracked combat vehicles from the U.S. military inventory, and permit deployment of a five-division force in one month rather than three.[27]

Some imagine going even further with more futuristic weapons. They envision capabilities such as intercontinental artillery, space-based weapons that could be rapidly unleashed at targets on earth only a few hundred kilometers below, and directed-energy weapons such as lasers.[28]

VULNERABILITY. The final major school of RMA thinking is motivated by worry as much as optimism or "technophilia." It highlights the growing threats posed by enemy cruise, antiship, and ballistic missiles; advanced satellite technologies for communications and targeting; sea mines and advanced diesel submarines; the physical and electronic vulnerabilities of information and communications systems on which the U.S. armed forces increasingly depend; the proliferation of chemical and biological weapons; and the enduring challenges of urban and infantry battle. These technologies could make it much harder for the United States to reach foreign ports safely, keep those ports as well as airfields and other infrastructure safe from enemy attack, and protect troops on the battlefield.[29]

26. Richard Danzig, "Report of the Secretary of the Navy," in William S. Cohen, *Annual Report to the President and the Congress* (Department of Defense, 1999), pp. 205–06.

27. Steven Lee Myers, "Army Is Restructuring with Brigades for Rapid Response," *New York Times*, October 13, 1999, p. A16; Robert Suro, "Chief Projects an Army on Wheels," *Washington Post*, October 13, 1999, p. A23.

28. On stealth bombers, see, for example, Perry, Pfaltzgraff, and Conway, *Long-Range Bombers*, and Barry M. Blechman and Paul N. Nagy, *U.S. Military Strategy in the 21st Century* (Arlington, Va.: IRIS Independent Research, 1997); on arsenal ships, see Andrew F. Krepinevich Jr., *A New Navy for a New Era* (Washington, D.C.: Center for Strategic and Budgetary Assessments, 1996); on long-range and space weapons, see Harlan Ullman and others, *Shock and Awe: Achieving Rapid Dominance* (National Defense University Press, 1996), and George Friedman and Meredith Friedman, *The Future of War: Power, Technology, and American World Dominance in the 21st Century* (Crown, 1996); for an army view, see Major General Edward G. Anderson III and Major Michael Linick, "Ensuring Future Victories through Land Power Dominance: The U.S. Army Modernization Strategy," *National Security Studies Quarterly*, vol. 2 (Autumn 1996), pp. 1–18.

29. Andrew F. Krepinevich Jr., *The Conflict Environment of 2016: A Scenario-Based Approach* (Washington, D.C.: Center for Strategic and Budgetary Assessments, 1996); General Charles C. Krulak, "Operational Maneuver from the Sea: Building a Marine Corps for the 21st Century," *National Security Studies Quarterly*, vol. 2 (Autumn 1996),

There appears to be ample reason for worry. At present, the United States is easily the world's best military force. It would be very fortunate if its dominance were to grow in the future; the opposite trend may be more likely due to the processes of technological diffusion and proliferation.[30] The vulnerability school of thought frequently invokes the term asymmetric warfare in arguing that future adversaries will choose to attack the United States differently than the United States would choose to fight them. That conclusion applies both to the battlefield, and to the American homeland, since foes might attempt terrorist acts against U.S. civilian and economic centers in an attempt to deter or defeat U.S. military action against them.[31]

OTHER RMA SCHOOLS OF THOUGHT. Others make even bolder claims about future warfare. Some assert that both the economy and the nature of warfare will change more in coming years than at any time since the industrial revolution and the age of Napoleon. That era saw the full human and material resources of the modern nation-state devoted to warfare for the first time; it therefore witnessed not only technical innovation, but also major transformation in the structure of militaries and their relation to the state. The best-known proponents of this "third wave" vision, which places the modern information revolution on historical par with the agricultural and industrial revolutions, are Alvin and Heidi Toffler.[32]

Other bold thinkers have posited possibilities such as essentially limitless energy sources, the complete transparency of the oceans, and other technological breakthroughs for which there is no current scientific basis. Slightly more restrained, but still extremely optimistic, visions of technological progress posit developments like a quintupling in the speed of battlefield maneuver between Desert Storm and 2010, from 40 to 200 kilometers per hour.[33]

---

pp. 19–23; Robert J. Bunker, *Five-Dimensional (Cyber) Warfighting: Can the Army after Next Be Defeated through Complex Concepts and Technologies?* (Carlisle Barracks, Pa.: U.S. Army War College Strategic Studies Institute, 1998); Blechman and Nagy, *U.S. Military Strategy in the 21st Century,* pp. 11–16, 68–70.

30. John Arquilla, "The 'Velvet' Revolution in Military Affairs," *World Policy Journal* (Winter 1997–98), p. 42.

31. See Gary Hart and others, *New World Coming: American Security in the 21st Century* (Alexandria, Va.: U.S. Commission on National Security/21st Century, September 1999).

32. Alvin Toffler and Heidi Toffler, *War and Anti-War: Survival at the Dawn of the 21st Century* (Boston: Little, Brown, 1993); Admiral Arthur K. Cebrowski and John J. Garstka, "Network-Centric Warfare: Its Origin and Future," *Proceedings* (U.S. Naval Institute, January 1998), pp. 29–35.

33. This is the view of one of the U.S. Army's leading proponents of the RMA. See Robert H. Scales Jr., "Cycles of War: Speed of Maneuver Will Be the Essential Ingredient of an Information-Age Army," *Armed Forces Journal International,* vol. 134 (July 1997), p. 38. On the trans-

To the extent that such visionaries participate in the actual RMA debate today—and many do not, since their time horizons are too long to be immediately relevant to most Pentagon decisionmaking—they can generally be lumped into the global reach, global power school. They envision futuristic warfare as far less dependent on large combat vehicles and, hence, far less constrained by geography and distance than has been the case in the twentieth century.

On the other extreme, another RMA school of thought might be defined to include those who believe that major changes in military affairs are coming but who do not yet claim to understand their implications. By comparison with the revolutionaries described previously, this school of thought is notable for its caution and patience. Its proponents find the contemporary RMA hypothesis appealing but remain undecided about its implications and unconvinced that it is near culmination. They are therefore wary of proposals for any radical makeover of the American military.[34] In this regard, they generally see more eye-to-eye with the system of systems than the DBK or global reach, global power schools. Their belief that the country should focus more attention and resources on defense activities such as experimentation is a point with which I concur, and which is revisited later in this book.

The Pentagon and Joint Vision 2010

Where does the Department of Defense (DoD) stand in the RMA debate? The answer is that, in terms of its rhetoric, it merges all of the aforementioned schools of thought into an aggregate position that reflects a grandiose and ambitious interpretation of what a contemporary revolution in military affairs should entail. In terms of its force structure, weapons programs, and budgets, however, there is little evidence that it is rapidly trying to transform its forces. To a large extent, the Pentagon is offering old wine in new bottles: a traditional set of budgetary preferences and priorities dressed up as a blueprint for a revolution in military affairs.

In 1997 the reports of the Pentagon's own Quadrennial Defense Review (QDR) and the independent National Defense Panel (NDP) gave strong sup-

parency of the oceans and limitless energy sources, see Zalmay Khalilzad and David Shlapak, with Ann Flanagan, "Overview of the Future Security Environment," in Zalmay Khalilzad and Ian O. Lesser, eds., *Sources of Conflict in the 21st Century* (Santa Monica, Calif.: RAND, 1998), pp. 35–36.

34. Cohen, Eisenstadt, and Bacevich, *Knives, Tanks, and Missiles*, pp. 7–10.

port to the RMA concept. Both reports used the phrase revolution in military affairs repeatedly, leaving no doubt that they accepted that an RMA is under way. The long-term vision documents of the military services and former chairman of the Joint Chiefs of Staff John Shalikashvili had already done the same, so 1997 was, in effect, the culmination of the RMA's acceptance in mainstream U.S. defense thinking.[35]

The Pentagon's official version of the RMA is, as noted, remarkable for its ambition. It focuses on information systems, sensors, new weapons concepts, much lighter and more deployable military vehicles, missile defenses, and other capabilities. The watchwords for effecting this transformation, employed earlier in the *Joint Vision 2010* document, begin with information superiority or dominance. They also include the terms dominant maneuver, precision engagement, full-dimensional protection, and focused logistics. Dominant maneuver and focused logistics imply light, agile, deployable main combat forces. Precision engagement conjures up images of very accurate and lethal long-range firepower. Full-dimensional protection suggests, among other things, highly effective missile defenses.[36]

There is a certain irony in the fact that the cautious Pentagon has become, rhetorically at least, one of the most enthusiastic proponents of the RMA concept. In fact, the same QDR report that espoused information superiority, dominant maneuver, precision engagement, full-dimensional protection, and focused logistics also reported an explicit decision not to pursue these goals too fast. As such, it adopted the rhetoric of the RMA movement without taking on its sense of urgency or imminence. To put it differently, even as it accepted the RMA hypothesis, the Department of Defense made few plans to reorganize its main combat units, increase their interdependence and jointness, or alter priorities within the weapons modernization program. Nor did it increase the total amount of resources devoted to acquisition accounts—the sum total of procurement with research, development, testing, and evaluation.[37] This position seems inherently self-contradictory: to believe a revolution is imminent, yet make only modest adjustments to force structure and weapons programs. In fairness to the Pentagon, however, that cautious approach, even if not internally consistent, may be preferable to adopting a wrongheaded and radical RMA agenda.

35. See William S. Cohen, *Report of the Quadrennial Defense Review* (Department of Defense, May 1997), pp. iv–v, 39–51; National Defense Panel, *Transforming Defense: National Security in the 21st Century* (Arlington, Va.: December 1997), pp. 5–8.

36. Cohen, *Report of the Quadrennial Defense Review*, pp. 39–41.

37. See, for example, M. Thomas Davis, "Warfighting Transformation: A Slow Moving Process," briefing slides, Northrop Grumman Analysis Center, Rosslyn, Va., August 25, 1998.

Observing the Pentagon's inertia, the National Defense Panel tried to push the military to do more and do it faster. Its 1997 report critiqued the QDR for not adopting a sufficiently ambitious reorientation of Pentagon priorities. But just what the NDP had in mind, by way of a roadmap for the purported revolution, was difficult to discern as well. It laid out little in the way of specifics about which programs and efforts should be accelerated and which should be cut to make resources available for new priorities.[38]

Past Revolutions in Military Affairs

Although the RMA movement has frequently been vague about what a contemporary revolution in military affairs should entail, several of its proponents, as well as other scholars, have done a good job of revisiting military history to gain perspective on past revolutions. Their observations are useful in understanding what makes an RMA possible and what scale of effort is generally necessary to take advantage of it.

A Brief Chronology of Revolutions in Military Affairs

Historically there have been at least three different kinds of revolutions in military affairs. Previous RMAs have sometimes resulted from 1) the development of a single dominant technology like the crossbow, gunpowder, or nuclear weapon; 2) a fundamental change in how the state organized its resources for security, as in the Napoleonic era; or 3) changes in the way military technologies were linked together, as in the development of the blitzkrieg doctrine or carrier airpower before and during World War II.[39] All of these revolutions were characterized by the same basic attribute: those countries that exploited them first achieved major battlefield advantages as they made the tactics and technologies of their enemies obsolete.

One survey of past RMAs, compiled by Andrew Krepinevich, identified ten revolutions over the past 700 years, beginning with the infantry revolution of the fourteenth century. At that time, the introduction of the six-foot yew longbow gave archers the ability to penetrate the armor of riders on horseback. Next was the revolution of sail and shot, beginning toward the end of the fourteenth century and lasting for the next couple hundred

---

38. Cohen, *Report of the Quadrennial Defense Review*, pp. 21–27; National Defense Panel, *Transforming Defense*, pp. 1–3.

39. Cooper, "Another View of the Revolution in Military Affairs," pp. 117–19.

years. Oar-driven galley boats were replaced by sailing ships. Over time they came to carry both troops and artillery. In the fifteenth century, the artillery revolution unfolded on both land and sea. It was made possible by improvements in metallurgy that reduced the costs of gun barrels, as well as the process of corning that tripled the energy per unit weight of gunpowder. The final two revolutions of this time period were the fortress revolution of the sixteenth century, in which lower and thicker walls together with various types of fort superstructure made defended areas virtually impregnable, and the gunpowder revolution of the sixteenth and seventeenth centuries, in which infantry units took full advantage of the same improved gunpowder that had already made artillery more effective by adding muskets to their standard arsenals.[40]

The next five revolutions, as Krepinevich defines them, took place within the space of about 150 years. The first was the Napoleonic revolution, partly a technical achievement (due to the standardization and mass production of weapons like artillery) and partly a societal achievement in its adoption of the large conscript army. The land warfare revolution of the nineteenth century followed and built on Napoleon's achievements, also taking advantage of railroads, the telegraph, rifled muskets and artillery, and eventually the machine gun. The naval revolution of the nineteenth and twentieth centuries brought metal-hulled ships with turbine engines and long-range guns to the fore. The interwar revolutions in mechanization, aviation, and information took advantage of rapidly improving technology in the internal combustion engine, aircraft, radio, and radar and developed new concepts of operations such as blitzkrieg, carrier warfare, and amphibious assault. Last on this list of previous RMAs is the nuclear weapons revolution.[41]

What made these revolutions happen? In some cases, a technical innovation was so important that it virtually transformed warfare single-handedly. More commonly, innovative militaries needed to devise new and complicated organizational concepts and tactics to make use of weapons that could not be easily integrated into existing formations or battle plans. In addition, their political leaders needed to provide the necessary push and the requisite resources to turn fledgling plans and technological prototypes into new military machines.

Krepinevich's chronology of military revolutions is concise and informative. But students of military history should not think that there is any

40. Andrew Krepinevich Jr., "Cavalry to Computer: The Pattern of Military Revolutions," *National Interest*, no. 37 (Fall 1994), pp. 31–33.

41. Krepinevich, "Cavalry to Computer," pp. 34–36.

widely agreed-upon way to think about RMAs of the past—or even what terminology to use to describe them. Some would add people's war or guerrilla war to the list.[42] Others would use an entirely different taxonomy, going back further in history and identifying other key turning points in the evolution of military hardware, tactics, and organizations.[43] For example, Jared Diamond explains the European conquest of civilizations like the Aztec and Inca less by reference to specific weapons than to Europe's more sophisticated and urbanized economy, the root causes of which were agricultural breakthroughs resulting from grains and domesticated animals imported from the Middle East.[44]

Nonetheless, with the exception of Alvin and Heidi Toffler, few are now arguing that current human affairs offer the potential for as radical a change in warfare as that brought on by the early economic development of Europe, the consolidation of the Westphalian state system, or the Napoleonic revolution. Any contemporary RMA is likely to be more akin to what happened between the world wars, or perhaps to the anticolonial and nationalistic movements of the post–World War II era, than to the Napoleonic or Westphalian periods. It will probably involve changes in military technology and tactics more than a reordering of all major human activity, including warfare. Thus the aforementioned chronology of past RMAs should provide relevant perspective on what any contemporary revolution in military affairs might involve.

Making RMAs Happen

How does a country go about exploiting—or creating—a revolution in military affairs? And how can countries fail to recognize RMAs and potentially become vulnerable as a result? History can offer guidance on these questions as well.

Military organizations tend to be conservative. Absent compelling international threats or agendas, they may display excessive caution. They may well prefer new weaponry, high-payoff offensive doctrines, and maximum resources—but that hardly guarantees that they will be innovative. Often new weapons will be grafted onto preferred and preexisting ways of warfight-

---

42. Williamson Murray, "Thinking about Revolutions in Military Affairs," *Joint Forces Quarterly* (Summer 1997), pp. 69–76.

43. Martin Van Creveld, *Technology and War: From 2000 B.C. to the Present* (Free Press, 1991), pp. 1–6.

44. Jared Diamond, *Guns, Germs, and Steel: The Fates of Human Societies* (W. W. Norton, 1997), pp. 67–81.

ing and doing business. The promise of new technologies will often be hard for them to grasp, because this is difficult to assess using existing paradigms and concepts of operations. Often it is necessary for strong civilian leadership to push the military to innovate; otherwise, innovation may have to await battlefield defeats or may be only incremental as a military tries new technologies in battle and gradually comes to see that they offer great promise. For states that have not been recently defeated, however, and that have not had the "luxury" of fighting a number of recent battles, these paths to innovation may be inadequate.[45]

It was largely the impatient and driven Adolf Hitler, who insisted that the German military innovate and gave rise to the concept of blitzkrieg. In contrast, civilians played a lesser role in driving military doctrine in France; as a result, the organizational tendencies of its large military organization produced inertia followed by defeat. Britain represented an intermediate case. Its civilian leaders did not override the air force's preference for an aggressive strategic bombardment doctrine, but they did manage to promote a new air defense concept that helped the country fend off German aerial attacks in the Battle of Britain.[46]

Nonetheless, rarely can civilians carry out a military revolution on their own; they usually need assistance from a cohort of officers within the mainstream of a military service who come to support and encourage innovation. This can be seen by comparing the successful interwar experience of the U.S. Navy in developing carrier airpower with the generally failed experience of the British navy. The latter had allowed its aviators to be siphoned off by the air force while, under Rear Admiral William Moffett, the former built up a committed cadre of pilots within the rank and file of the U.S. Navy who championed the cause of naval airpower. This ensured that aircraft carriers would be vigorously researched and solidly funded during the critical years when their potential usefulness in war was suspected but not yet established.[47]

These examples underscore another important point for the contemporary RMA debate: that peacetime eras are often the best times for militaries to innovate. They can focus on new ideas and develop them more easily when war does not distract them, discourage risk-taking, or limit the scope of innovation by placing severe time pressure on the development of new

45. See Barry R. Posen, *The Sources of Military Doctrine: France, Britain, and Germany between the World Wars* (Cornell University Press, 1984), pp. 41–80.

46. Posen, *Sources of Military Doctrine*, pp. 220–41.

47. Stephen Peter Rosen, *Winning the Next War* (Cornell University Press, 1991), pp. 13–18, 76–80, 96–100.

warfighting concepts and weaponry. That innovation can occur during long wars does not mean that one should wait for actual combat to serve as the test bed for invention.[48]

Why the RMA Debate Is Important

If the hypothesis that an RMA was attainable or under way simply amounted to a prediction, it would matter little whether its proponents were right. We could simply wait to find out and view the subject as intellectually interesting but not particularly important.

However, as RMA proponents frequently claim and as the previous historical discussion shows, military revolutions are the purposeful creations of people. They are created by a combination of technological breakthrough, institutional adaptation, and warfighting innovation.[49] They are not emergent properties that result accidentally or unconsciously from a cumulative process of technological invention.

For this reason, the RMA debate matters. If the proponents of the RMA hypothesis are right, it could be dangerous for the United States not to heed their counsel.[50] By contrast, if they are wrong, it could be harmful to the country's security interests—or, at a minimum, wasteful of resources—to adopt their recommendations.

RMA proponents tend to argue that more budgetary resources should be devoted to innovation—research and development (R&D), procurement of new hardware, frequent experiments with new technology—and less money to military operations, training, and readiness.[51] To free up funds for an RMA transformation strategy, some would reduce U.S. global engage-

48. Rosen, *Winning the Next War*, pp. 57–105, 251–61.

49. For one such view, see Keaney and Cohen, *Gulf War Air Power Survey Summary Report*, pp. 235–51.

50. For a similar view, see Jeremy Shapiro, "Information and War: Is It a Revolution?" in Zalmay M. Khalilzad and John P. White, eds., *The Changing Role of Information in Warfare* (Santa Monica, Calif.: RAND, 1999), pp. 114–17.

51. Some might expect that, given the country's fiscal good fortune, surpluses will allow overall defense spending to go up considerably—making possible an RMA research, experimentation, and investment agenda without requiring cutbacks in current force structure or operations. However, that is an optimistic, and in the end unlikely, assessment. After all, even under President Clinton's widely advertised defense spending initiative of January 1999, real defense spending is only to be about $5 billion more in 2004 than in 1999. Republicans, for their part, would cut defense spending considerably after 2004 under their tax-cut proposal of 1999. Congressional Budget Office, *An Analysis of the President's Budgetary Proposals for Fiscal Year 2000* (1999), p. 46.

ment and weaken the military's deterrent posture.[52] For example, in its 1997 report, the National Defense Panel dismissed the current two-war strategy as obsolete (without suggesting what should replace it, however). The NDP also suggested that U.S. military retrenchment from forward presence and peacekeeping operations might be needed simply to free up money to promote the so-called revolution in military affairs. These suggestions, if adopted, would have important effects on U.S. security policy; they should not be accepted simply on the basis of vague impressions that an RMA may be achievable.[53]

To use the lexicon of the Pentagon's current official strategy, articulated in Secretary of Defense William Cohen's 1997 Quadrennial Defense Review, RMA proponents would put more resources into preparing U.S. forces for an uncertain future. Assuming a given Pentagon budget level, they would therefore put less emphasis on shaping the future military environment through American military engagement overseas or being ready to respond to near-term challenges to U.S. interests by the likes of Saddam Hussein and Kim Jong-Il. To use the terminology made popular by the NDP, the alleged RMA is also sometimes described as a transformation—implying a more managed and controlled process, but still a radical one.[54]

The notion of an RMA has other implications for security policy. Would a radical transformation of the U.S. military make American forces too complex to interoperate with those of allied countries in future multinational military operations? Would an RMA dramatically reduce the need for overseas military bases for the American armed forces by making it possible to project decisive power rapidly from the North American homeland? Could RMA technologies transform the arms control agenda, perhaps making it possible to consider the abolition of nuclear weapons? These types of ques-

---

52. For an argument in favor of taking a large part of the active force structure "off line" so as to devote it to experimentation and acceleration of the RMA, see James R. Blaker, "The American RMA Force: An Alternative to the QDR," *Strategic Review*, vol. 25 (Summer 1997), pp. 21–30; for a similar but more general argument, see also Richard K. Betts, *Military Readiness: Concepts, Choices, Consequences* (Brookings, 1995), pp. 35–84. For the view of a conservative critic of the RMA concept, see Frederick W. Kagan, "Wishful Thinking on War," *Weekly Standard*, December 15, 1997, pp. 27–29. Kagan argues that the country may need to spend more on technology—but must not do so at the expense of its present engagement and deterrence strategies.

53. National Defense Panel, *Transforming Defense*, pp. vii, 2, 23, 49, 59, 79–86.

54. The National Defense Panel used the terms revolution in military affairs and transforming defense repeatedly and interchangeably, however, so it is not clear that there is really any distinction between them in the current debate. See National Defense Panel, *Transforming Defense*; see also James R. Blaker, "Revolution(s) in Military Affairs: Why the Critique?" *National Security Studies Quarterly*, vol. 5 (Winter 1999), p. 86.

tions are being raised in the RMA debate; their potential implications are far from academic.

Given its technical nature, should the RMA debate be left to just the Pentagon, military weapons laboratories, and the defense industry? The answer is clearly no. Pursuit of an RMA vision requires trade-offs with other types of U.S. security policy interests and budgetary priorities; the issue is not solely a technical military one.

Nor, for all its classified databases on weapons capabilities and performance, does the Pentagon have a monopoly on good defense analysis. The DoD's elaborate models reportedly did a worse job of predicting casualties in Desert Storm than did most similar (though less complex) studies by independent experts.[55] Many of DoD's models remain somewhat antiquated and are generally unable to assess the roles played by individual weapons systems. That fact poses a major problem when considering new technologies: if they cannot be properly inserted into the models and their individual consequences studied, the Pentagon's most rigorous methods are of limited use.[56] Moreover, much of the modeling and simulation work done by DoD is sponsored by the individual military services. In the real world of Pentagon politics, that fact immediately makes the findings of a study done by one service suspect among the others.

There is clearly an important role for the Pentagon in testing the RMA hypothesis and pursuing future technological visions for the U.S. armed forces. But the broader policy community must weigh in, too.

Reasons to Be Wary of the RMA Hypothesis

Given the large opportunity costs of pursuing an RMA agenda, a certain wariness about the concept is appropriate. There are other reasons to be skeptical as well.

55. For a summary of the various predictions made at the time, see Michael Miller and others, "Costs of Operation Desert Shield" (Congressional Budget Office, January 1991), p. 15.

56. See Khalilzad and Ochmanek, "Rethinking U.S. Defense Planning," pp. 43–64; National Defense Panel, "Assessment of the May 1997 Quadrennial Defense Review" (Arlington, Va.: May 15, 1997), p. 8; General Accounting Office, *Quadrennial Defense Review: Opportunities to Improve the Next Review*, GAO/NSIAD-98-155 (June 1998), pp. 5, 25, 28; Robert P. Haffa, "Planning U.S. Forces to Fight Two Wars: Right Number, Wrong Forces," *Strategic Review* (Winter 1999), pp. 15–21; Defense Science Board, *Report of the Defense Science Board Task Force on Deep Attack Weapons Mix Study (DAWMS) Phase II* (Office of the Under Secretary of Defense for Acquisition and Technology, 1998).

History provides certain grounds for caution. Most contemporary RMA enthusiasts make reference to the interwar years and claim that we are in another period of similar potential, promise, and peril today. However, military technology advanced steadily and impressively throughout the twentieth century, including its latter half. Helicopters radically reshaped many battlefield operations after World War II. Satellite communications were first used militarily in 1965 in Vietnam, where aircraft-delivered, precision-guided munitions also made their debut in the early 1970s. Air defense and antitank missiles played major roles in the 1973 Arab-Israeli War. Stealth fighters were designed in the late 1970s.[57] Infrared sensors and night-vision technologies made their debut in this period as well.

It is far from obvious that military technology is now poised to advance even more quickly than it has in the last half century. Yet RMA proponents assert that it will when they call for a radical transformation strategy for current U.S. armed forces. No such radical DoD-wide transformation strategies were necessary to bring satellites, stealth, precision-guided munitions, advanced jet engines, night-vision equipment, or other remarkable new capabilities into the force in past decades.[58]

A cursory examination of the major U.S. conflicts of the 1990s also raises questions about the pace of change in military technology and capabilities. Notably, the first conflict of the decade was the most successful—not necessarily what one would expect if technology and tactics were evolving rapidly, as RMA believers assert. Desert Storm in 1991 was a far more convincing use of force than U.S. military operations in Somalia in 1993. It was also more impressive than the three-day bombing campaign against Iraq in late 1998, when U.S. and British airpower were unable to make any dent in Iraqi weapons of mass destruction capabilities, and set back Saddam's missile-making capabilities by no more than a year. That campaign underscored the simple fact that no matter how good U.S. weapons become—and they were better than in 1991, featuring more accurate cruise missiles as well as bunker-busting bombs—the United States cannot attack enemy targets that it cannot locate.[59]

57. Lawrence Freedman, *The Revolution in Strategic Affairs*, Adelphi Paper 318, International Institute for Strategic Studies (Oxford University Press, 1998), p. 21.

58. Van Creveld, *Technology and War*. Trevor Dupuy uses yet another categorization scheme—different from those of Krepinevich, Van Creveld, and others—to understand the history of military innovation. He groups all progress since 1800 together under the title of "the age of technological change." See Trevor N. Dupuy, *The Evolution of Weapons and Warfare* (Fairfax, Va.: Hero Books, 1984).

59. Steven Lee Myers, "In Intense but Little-Noticed Fight, Allies Have Bombed Iraq All

Desert Storm was also at least as impressive as NATO's seventy-eight-day air campaign against Serbia in 1999. Operation Allied Force, as the war over Kosovo was known, did showcase several new weapons that represented improvements on what was available in 1991, such as the so-called joint direct attack munition (JDAM) bomb dropped from the B-2 bomber with its highly accurate radar, and NATO's fairly large collection of UAVs. But even these systems did not permit NATO to attack Serbian armored forces in Kosovo in bad weather or to distinguish military vehicles from civilian ones without the direct intervention of a pilot's eye. They constituted a relatively modest set of advances given that, since Desert Storm, the U.S. military had had eight years to improve its capabilities—and spent an annual average of $100 billion on weapons acquisition to do so. Even if RMA proponents would have followed different investment strategies in the 1990s than did the Pentagon, the fact remains that substantial amounts of resources were devoted to areas such as advanced munitions as well as reconnaissance platforms like UAVs and joint surveillance and target attack radar system (JSTARS) aircraft—priorities that most RMA proponents would support. In truly revolutionary times, the United States might have been expected to have more to show for that $800 billion.[60]

RMA proponents are certainly right to believe that a successful military must always be changing. But the post–World War II U.S. military has already taken that adage to heart. The status quo in defense circles does not mean standing still; it means taking a balanced approach to modernization that has served the country remarkably well for decades. Indeed it brought on the very technologies displayed in Desert Storm that have given rise to the belief that an RMA may be under way.[61] It is not clear that we need to accelerate the pace of innovation now.

Year," *New York Times*, August 13, 1999, p. A1; "Assessing the Bomb Damage," *Wall Street Journal*, December 22, 1998, p. A18; Robert Wall, "New Weapons Debut in Attacks on Iraq," *Aviation Week and Space Technology*, December 28, 1998, pp. 14–15.

60. In fiscal years 1991 through 1999, the Pentagon spent about $935 billion on procurement plus research, development, testing, and evaluation, so something more than $800 billion was spent between the end of Desert Storm (roughly halfway through fiscal year 1991) and the beginning of Operation Allied Force (about halfway through fiscal year 1999). See Office of the Under Secretary of Defense (Comptroller), *National Defense Budget Estimates for FY 2000*, pp. 128–29.

61. For sound warnings about both dismissing the RMA and jumping on the bandwagon too enthusiastically, see Colin S. Gray, *The American Revolution in Military Affairs: An Interim Assessment* (Camberley, England: Strategic and Combat Studies Institute, 1997), pp. 5–7, 33–34; for a reminder that militaries must always be innovating and changing, see Jonathan Shimshoni, "Technology, Military Advantage, and World War I: A Case for Military Entrepreneurship," *International Security*, vol. 15 (Winter 1990), pp. 213–15.

Moreover, radical innovation is not always good. If the wrong ideas are adopted, transforming a force or its tactics can cause harm. For example, in the world wars, militaries overestimated the likely effects of artillery as well as aerial and battleship bombardment against prepared defensive positions, meaning that their infantry forces proved much more vulnerable than expected when they assaulted enemy lines.[62] Britain's radically new all-tank units were inflexible, making them less successful than Germany's integrated mechanized divisions in World War II. Strategic aerial bombardment did not achieve nearly the results that had been expected of it; airpower was much more effective as close-air support for armored formations in blitzkrieg operations.[63] Later on, the U.S. Army's Pentomic division concept, intended to employ tactical nuclear weapons, was adopted for a time but abandoned in 1961.[64]

Despite their haste to push the revolution along, RMA promoters tend to lack clear and specific proposals for how to do so. In that light, even if they are right that an RMA may be within reach sometime in the foreseeable future, they may be quite wrong about what should be done about it in the near future. In practical terms, there is a major distinction between the early stages of a possible RMA and the later stages. As Stephen Rosen writes: "The general lesson for students or advocates of innovation may well be that it is wrong to focus on budgets when trying to understand or promote innovation. Bringing innovations to fruition will often be expensive. Aircraft carriers, fleets of helicopters, and ICBM forces were not cheap. But *initiating* an innovation and bringing it to the point where it provides a strategically useful option has been accomplished when money was tight ... Rather than money, talented military personnel, time, and information have been the key resources for innovation."[65]

62. John Keegan, *The First World War* (Alfred A. Knopf, 1999), p. 20; Dan Goure, "Is There a Military-Technical Revolution in America's Future?" *Washington Quarterly* (Autumn 1993), p. 185; Dupuy, *The Evolution of Weapons and Warfare*, pp. 218–20, 258–66.

63. Pape, *Bombing to Win*, pp. 87–136, 254–313; Brian Bond and Williamson Murray, "British Armed Forces, 1918–1939," in Allan R. Millet and Williamson Murray, eds., *Military Effectiveness*, vol. 2 (Boston: Unwin Hyman, 1988).

64. Stephen Biddle, "The Past as Prologue: Assessing Theories of Future Warfare," *Security Studies*, vol. 8 (Autumn 1998), pp. 1–74. Andrew J. Bacevich, *The Pentomic Era: The U.S. Army between Korea and Vietnam* (National Defense University Press, 1986); John Keegan, *A History of Warfare* (Alfred A. Knopf, 1993), pp. 362–79; Van Creveld, *Technology and War* (Free Press, 1991), pp. 193–95; Rosen, *Winning the Next War*, pp. 13–18, 37–38.

65. Rosen, *Winning the Next War*, p. 252.

Some RMA proponents—notably those of the vulnerability school—are convinced that the United States must find new ways to mitigate risks to its forces and look to a contemporary RMA for the answers. They are probably right that future adversaries will make greater use of sea mines, cruise missiles, possibly chemical weapons, and other means to attempt to deny the U.S. military the ability to build up forces and operate from large, fixed infrastructures as in Desert Storm. But they may not be right to suggest, as a number do, that the solutions to these problems are in the realm of advanced weaponry. True, long-range strike platforms, missile defenses, short-takeoff aircraft, and other such advanced technologies may be part of the solution. But so might more minesweepers, roll-on/roll-off transport vessels, concrete bunkers, and other relatively low-tech solutions to protecting, hardening, and dispersing supplies and infrastructure. The military services already are biased in favor of procuring advanced weaponry at the expense of equally important but less advanced hardware. By emphasizing modernistic and futuristic technology, the RMA movement could reinforce this existing tendency, quite possibly to the nation's detriment.

Most centrally, one should be skeptical about the revolution in military affairs hypothesis because many of its key technical underpinnings have not been well established and may not be valid. Proponents of the RMA concept often make passing mention of Moore's "law"—the trend for the number of transistors that can fit on a semiconductor chip to double every eighteen to twenty-four months—and then extrapolate such a radical rate of progress to much different realms of technology. For example, in its 1997 report the National Defense Panel wrote: "The rapid rate of new and improved technologies—a new cycle about every eighteen months—is a defining characteristic of this era of change and will have an indelible influence on new strategies, operational concepts, and tactics that our military employs."[66] However, conflating progress in computers with progress in other major areas of technology is unjustified. To the extent RMA believers hinge most of their argument on advances in modern electronics and computers, they are at least proceeding from a solid foundation. When they expect comparably radical progress in land vehicles, ships, aircraft, rockets, explosives, and energy sources—as many do, either explicitly or implicitly—they are mistaken, as is shown in subsequent chapters.

---

66. National Defense Panel, *Transforming Defense*, pp. 7–8.

Conclusion

This chapter is inconclusive. On the one hand, we know that revolutions in military affairs do happen and that their implications for warfighting and international security can be dramatic. At the same time, institutional and political factors often impede their adoption. Given these historical realities, the United States must be vigilant; otherwise, out of complacency or inertia, it could fail to exploit new technological and warfighting opportunities, allowing adversaries to obtain them first. Moreover, both civilians and uniformed military officers should think hard about the potential for military innovation, since history suggests that both groups have important roles to play in achieving successful advances.

On the other hand, if the technological moment is not ripe or if a new concept of warfighting operations is flawed, some attempted RMAs wind up making things worse. Many new technologies, as well as associated tactics and operational concepts, can be gradually integrated into a force without radically transforming it. Wholesale restructuring is often not necessary or even desirable. Even if an RMA is attainable, one must think carefully about how to pursue it programatically. Major and expensive weapons procurement programs have their place, but not at all times. At many stages in a defense transformation process the key to success is patient, gradual, and rather inexpensive research and experimentation. Embarking on an ambitious RMA transformation program could also have large opportunity costs for the United States and its foreign policy by weakening efforts such as overseas deterrence, peacekeeping and conflict resolution, or other forms of global engagement.

In short, neither a historical review of previous revolutions in military affairs, nor casual allusions to current trends in electronics and computer technologies, can determine if an RMA is attainable in the early years of the twenty-first century. Nor can they tell us how to realize that RMA even if it is within reach. To answer those questions, it is necessary to conduct a more detailed examination of trends in contemporary technology and their implications for warfare. This is the task to which the rest of this book is devoted.

CHAPTER THREE

Sensors, Computers, and Communications

Much of the debate over the purported revolution in military affairs (RMA) lacks serious attention to technical issues. Certain facts—notably the rapid advances in computing speed and the performance of stealth technologies and precision munitions in the 1991 Persian Gulf War—are frequently invoked. But other trends in defense technology tend to be ignored or cited only anecdoctally—and often incorrectly.

This is understandable. Attempting even a rough and imperfect summary of what key defense technologies can now do, and what they are likely to be able to do in a decade or two, is a daunting task. However, it must be attempted. Implicitly, all participants in the RMA debate are operating under assumptions about the march of technology.[1] That they do not describe their views, or provide much evidence for them, does not change this basic fact. But important assumptions about technological trends and innovations should be made explicit, not implicit, and debated thoroughly. They are critical to the entire RMA debate.

In that spirit, this chapter and the next provide an overview of current military technology as well as a projection of likely future trends and possibilities. Rather than focusing on technologies where progress is fastest, such as computers, these chapters survey most major areas of military technology—and underscore what will probably or certainly *not* be possible as much as what will. This approach does not definitively invalidate the notion

---

1. Joshua Epstein has made a similar, and very convincing, argument about the need for models in assessing military balances—and about the fact that virtually everyone is using one whether they acknowledge the fact or not. See Joshua M. Epstein, "Dynamic Analysis and the Conventional Balance in Europe," *International Security*, vol. 12 (Spring 1988), pp. 158–59.

that a revolution in military affairs is under way. But it does throw cold water on some of the more expansive technological visions espoused by RMA supporters. It also reveals realms of warfare in which any RMA will likely have only modest implications. Specifically, largely because of constraints on the performance of future battlefield sensors, infantry and urban warfare are likely to change much less than armored and naval combat. Even within the latter realms, high-technology forces may have serious vulnerabilities and limitations.

The time frame of the analysis is 2020—much longer than the Pentagon's typical six-year time horizon for planning and budgeting, but actually not that distant a year in the scheme of defense technology. As of the year 2000, for example, the United States continues to operate combat aircraft almost all of which were developed in the 1970s. Even in areas of munitions and communications systems, much of what is likely to be available in 2020 will generally need to be researched and developed fairly soon. So my chronological scope, while admittedly ambitious, stops well short of science fiction and futurism.

Two main methodologies guided my efforts: first, a survey of technical literature, including defense periodicals and scientific journals, and second, basic principles of physics. After having reached preliminary results with these approaches, I then presented my findings before a number of weapons scientists and technology experts to obtain their critiques and suggestions.

This chapter focuses on sensors as well as communications systems and computers. Chapter 4 addresses technologies that deploy troops, move them around the battlefield, and make them lethal to an enemy. The two chapters are hardly comprehensive, but they do cover most of the key technologies that will determine the overall thrust of future military innovation and performance.

This chapter makes particularly frequent use of the laws of physics to illustrate innate limitations on key military technologies. It is important to realize that this approach may actually *overstate* likely technological progress over the next two decades. Clearly, not everything that is theoretically possible will be practically achievable within any specific length of time. Moreover, even when technologies are invented and developed, they may not be immediately put to good military use. In that sense, this chapter tends to establish an upper bound on technological innovation between 2000 and 2020. It is only prudent for the United States to assume that its potential adversaries may achieve what is theoretically possible. But it should not underestimate the difficulty that it—or, even more likely, other countries—will face in actually pushing the limits of the technologically feasible.

Sensors

Sensor technologies can be grouped into three categories: those relying on visible and near-visible (infrared and ultraviolet) light; radar and radio waves; and all other types, most of which rely on detecting motion of the air, ground, or water. Nuclear, biological, and chemical detectors are also included in this last category.

Visible and Near-Visible Light Sensors

Sensors in this category exploit visible, infrared (IR), and ultraviolet (UV) light—all various types of electromagnetic radiation with fairly similar wavelengths. For the most part, the sensors are passive, simply detecting radiation emitted or reflected by the object at issue. Laser sensors are an exception to this generality, however. They often emit beams of light and then search for any reflections—a technique called lidar (light detection and ranging), that, as its name suggests, is akin to radar (radio detection and ranging).

Optical and infrared sensors have remarkable capabilities today. For example, the telescope on the KH-11 spy satellite orbiting 150 miles over earth can resolve objects less than six inches across. As of mid-1998, the United States reportedly operated three optical imaging satellites of KH-11 or later vintages.[2] Even commercial satellites now often possess resolution capabilities of one to two meters. Sensors on aircraft, which operate at considerably lower altitudes, can do even better. For example, unmanned aerial vehicles (UAVs) within ten kilometers of an object can resolve to ten centimeters, and those within one kilometer to just one centimeter.[3]

Infrared detectors exceed the precision of radar in some tactical settings; for example, detecting and identifying aircraft out to ranges of greater than ten kilometers.[4] Current infrared detectors make it possible to form images

2. Jeffrey T. Richelson, *America's Secret Eyes in Space: The U.S. Keyhole Spy Satellite Program* (Harper and Row, 1990), pp. 186–87, 362; Craig Covault, "Eavesdropping Satellite Parked over Crisis Zone," *Aviation Week and Space Technology*, May 18, 1998, p. 30.

3. Martin C. Libicki, "DBK and Its Consequences," in Stuart E. Johnson and Martin C. Libicki, eds., *Dominant Battlespace Knowledge* (National Defense University, 1996), p. 25; Ann Florini, "The End of Secrecy," *Foreign Policy*, no. 111 (Summer 1998), p. 55; Joseph C. Anselmo, "Imagery Satellite Costs Prompt NRO Delay," *Aviation Week and Space Technology*, May 25, 1998, p. 24.

4. More than 500 U.S. combat aircraft are now equipped with infrared devices for navigational and targeting purposes. Mark Hewish and Joris Janssen Lok, "Passive Target Detection for Air Combat Gathers Pace," *Jane's International Defense Review*, vol. 32 (February 1998), pp. 32–37.

Figure 3-1. *Reflectivity of U.S. and Soviet Camouflage and Healthy Green Vegetation*

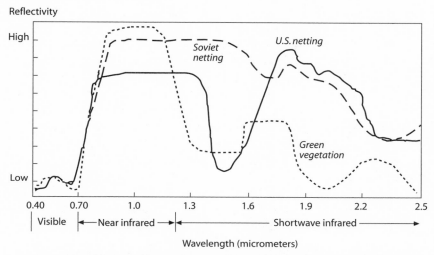

Source: "Growing Intelligence Operation Focuses on New Types of Signals," *Aviation Week and Space Technology*, August 2, 1999, p. 54 (reproduced from Office of MASINT, *The Multispectral Users Guide* [Department of Defense]).

of an entire aircraft or tank, rather than relying on tracking engine exhaust. This is a trend likely to produce benefits, but also unfortunate implications, for the United States. As they are used more widely, these advanced infrared detectors will greatly complicate the efforts of aircraft and tanks to use flares to lead missiles off their targets (since the more advanced devices look not only for heat but also for a target of a certain size and shape). Already, infrared seekers on man-portable surface-to-air missiles like the U.S. Stinger and Russian Igla-1 (SA-16) cover a wider range of frequencies, often including ultraviolet as well as infrared bands, that make conventional heat decoys far less effective in spoofing them.[5] As shown in figure 3-1, such seekers also make it harder to hide military equipment, since camouflage that can mimic vegetation at one frequency or wavelength may not fool the sensors at another.

5. Steven J. Zaloga, "Air-Defense Missiles: Recent Trends in the Threat," *Journal of Electronic Defense,* vol. 21 (November 1998), pp. 37–44; Charles R. Stumpfel, "Imaging In-Flight Projectiles by Infrared Emission," paper prepared for the Test Technology Symposium '97, Johns Hopkins University Applied Physics Laboratory, Laurel, Md., March 1997, available at (www.tecom.army.mil/tts/1997/proceed/stumpfel.html [September 1999]); Alan Vick and others, *Enhancing Air Power's Contribution against Light Infantry Targets* (Santa Monica, Calif.: RAND, June 1996), pp. 70–75.

Night-vision detectors for infantrymen now work well out to one to two kilometers, meaning that soldiers can now see as far as most of their weapons can accurately fire.[6] These devices are often based on simply intensifying the faint visible light that an object may reflect from the moon, city lights, or other sources. More advanced devices, as well as the detectors on platforms like modern tanks, typically rely on thermal imagers—that is, on infrared detectors. These devices pick up the heat released at infrared wavelengths by various objects, so they work even in perfectly dark conditions.[7]

Improvements in cooling the elements in IR sensors that distinguish temperatures, and other developments, will allow further increases in detection range. This means weak signals passing through a light rain that would be undetectable today may be able to be picked up in the future. The range of some IR systems will increase by 25 to 50 percent as a result of such upgrades.[8] For example, the LANTIRN infrared navigation and targeting system is being improved so that aircraft may fly as high as 40,000 feet (in contrast to the current 25,000-foot ceiling) while using it effectively.[9] In some situations, infrared sensors operating in a search mode may be able to locate snipers camouflaged from visual observation by picking up their heat signatures or those of the weapons and rounds they fire.[10]

It may also be possible to use lidar systems to search for snipers by locating their optics or to find other types of targets.[11] For example, Los Alamos National Laboratories is developing, in conjunction with the U.S. Army, a lidar system to detect and track aerosols characteristic of biological weapons use from as far away as thirty kilometers. This technology will help troops steer clear of suspicious clouds or send in equipment to sample and identify their composition.[12] Lasers may be useful for many other missions as

6. Greg Caires, "Reimer: Army Divisions More Lethal, Capable than Five Years Ago," *Defense Daily*, March 5, 1997, p. 340.

7. Mark Hewish, "Lifting the Veil of Darkness," *Jane's International Defense Review,* vol. 30 (June 1997), pp. 49–56.

8. Vick and others, *Enhancing Air Power's Contribution against Light Infantry Targets,* pp. 23–24, 34.

9. Bryan Bender, "USA Ponders LANTIRN Lessons after Iraq Raids," *Jane's Defence Weekly,* January 20, 1999, p. 8.

10. Vick and others, *Enhancing Air Power's Contribution against Light Infantry Targets,* p. 56; Mark Hewish and Rupert Pengelley, "Snipers Fire at Their Peril," *Jane's International Defense Review,* vol. 30 (July 1997), pp. 30–36.

11. Giovanni de Briganti, "Aerospatiale Unit Targets Exports for Bosnia-Tested Sniper Sensor," *Defense News,* September 15–21, 1997, p. 46; Libicki, "DBK and Its Consequences," p. 26.

12. Presentation by Ken McKenna, "Combating Chemical/Biological Attacks," Los Alamos National Laboratories, Los Alamos, N.M., April 6, 1998; John G. Roos, "WMD Defense:

well, such as measuring wind speeds to allow precision munitions to make minute changes in their final trajectories to improve accuracy, or setting up monitoring "fences" that would allow detection of the wakes of stealthy cruise missiles as they passed by.[13]

As robotics improve, existing sensor technologies will attain new capabilities simply by being able to go where they now cannot. Devices like vertical-lift UAVs, Matchbox toy–size wheeled vehicles, and six-inch-diameter aerial vehicles (perhaps powered by tiny jet engines or fired out of artillery tubes) may be developed to transport sensors around a battlefield.[14] Miniature video cameras with 72,000 pixels are already available and are soon expected to improve their resolution tenfold.[15] Some of these small devices could operate as networks, allowing for the fact that not all will survive their mission and that data will have to be pooled to paint a complete picture of a given building or area. Others might be dedicated to individual troops or squads, providing them reconnaissance assets under their own direct control. The chief technological challenges facing microrobotics are packing enough power supply into an object of less than one-foot diameter to allow it to move about and developing the necessary software. There appears to be a good chance that both hurdles can be overcome in the next decade or so.[16] (Whether power requirements can be achieved at the level of nanotechnology, where objects measure only a few atoms or molecules across, is a different matter. This is more a question of biology and chem-

Sensor Proliferation and Networking Are Keys to Future Chemical and Biological Defense Programs," *Armed Forces Journal International*, vol. 135 (May 1998), pp. 36–41; David Mulholland, "Chem-Bio Attack Fears Drive Detection Device Market," *Defense News*, April 20–26, 1998, p. 46; Mark Hewish, "On Alert against the Bio Agents," *Jane's International Defense Review*, vol. 31 (November 1998), p. 54; Robert F. Service, "Coming Soon: The Pocket DNA Sequencer," *Science*, vol. 282 (October 16, 1998), pp. 399–401; and Director, Defense Research and Engineering, *Joint Warfighting Science and Technology Plan* (Department of Defense, 1997), p. IV-I-5.

13. William B. Scott, "'Beam' Weapons Edging into Arsenal," *Aviation Week and Space Technology*, July 5, 1999, pp. 53–55.

14. Michael A. Dornheim, "Several Micro Air Vehicles in Flight Test Programs," *Aviation Week and Space Technology*, July 12, 1999, p. 47; George I. Seffers and Mark Walsh, "DoD Wants 6-Inch Planes to Fly Stealthy Operations," *Defense News*, November 10–16, 1997, p. 3; William B. Scott, "Gun-Fired 'WASP' Locates Targets," *Aviation Week and Space Technology*, August 31, 1998, p. 57; Robert Wall, "Darpa Eyes Helo UAV," *Aviation Week and Space Technology*, June 8, 1998, p. 24.

15. "A Personal Eye in the Sky," *Economist*, January 9, 1999, pp. 73–74.

16. George I. Seffers, "Special Operations Forces Want to Deploy with Robots," *Defense News*, April 13–19, 1998, p. 3; Michael A. Dornheim, "Tiny Drones May Be Soldier's New Tool," *Aviation Week and Space Technology*, June 8, 1998, pp. 42–48.

istry than traditional engineering. Regardless, objects of this size would be unable to transport most types of sensors.[17])

Radar and Radio Emission Sensors

Among the most critical military sensors are those operating in a part of the electromagnetic spectrum with wavelengths roughly 1 million times longer than light—the province of radar and radio.[18] Radar and radio are strongly related; the concept of radar was in fact developed when radio signals were found to be blocked by passing objects, such as ships.[19]

Radar has evolved greatly since its invention in the early twentieth century. For example, radar now often operates not by emitting out of a single rotating waveguide, but by electronically steering beams formed by individual emissions from thousands of small devices on a fixed surface. These instruments are known as phased-array radars. They also have the ability to selectively tune out emissions that may be coming at them from certain directions, drastically reducing their vulnerability to jamming. Among their most impressive manifestations involve the early detection and tracking of ballistic missiles and warheads, as performed by the U.S. Pave Paws, Cobra Dane, and PARCS radars; these sensors can pick up objects out to ranges of thousands of kilometers and can track hundreds at a time. Phased-arrays on Aegis cruisers and destroyers can also track hundreds of objects at a time. Lacrosse imaging satellites provide twenty-four-hour all-weather reconnaissance capabilities from low-earth orbit and have resolutions on the order of one meter.[20]

17. Robert F. Service, "Borrowing from Biology to Power the Petite," *Science*, January 1, 1999, pp. 27–28.

18. More specifically, radar usually operates at wavelengths between roughly one centimeter—corresponding to 30 GHz (gigahertz), that is, 30 billion cycles per second—and several meters (at 1.5 meters' wavelength, the frequency would be 200 million hertz, or MHz). Radio transmissions usually take place at longer wavelengths and thus lower frequencies, with corresponding wavelengths anywhere from tens of meters to thousands of kilometers. See Merrill I. Skolnik, *Introduction to Radar Systems*, 2d ed. (McGraw-Hill, 1980), p. 7.

19. See Steve Vogel, "Navy Lab Uncloaks a Secret, Celebrates Its Breakthroughs," *Washington Post*, June 19, 1998, p. A23.

20. Eli Brookner, "Phased-Array Radars," *Scientific American*, February 1985, pp. 94–103; Eli Brookner, *Radar Technology* (Dedham, Mass.: Artech House, 1978), p. 34; Bruce G. Blair, *Strategic Command and Control: Redefining the Nuclear Threat* (Brookings, 1985), pp. 224, 255; Commander Robert J. Engel, "Bringing Aegis to the Littorals," *Proceedings,* vol. 124 (January 1998), pp. 42–43; Captain Dan Meyer and Captain John Geary, "Aegis Computing Enters the 21st Century," *Proceedings* (January 1998), pp. 39–41; "AWACS Fitted with Phased Array Antenna," *Aviation Week and Space Technology*, June 8, 1998, p. 24.

One of the U.S. military's most notable recent innovations is the joint surveillance and target attack radar system (JSTARS) aircraft, first used in Desert Storm, that allows discrimination between objects like trucks and tanks as well as between moving and nonmoving vehicles.[21] It provided targeting information for a number of the aerial attacks in Operation Desert Storm even though its capabilities to distinguish between wheeled and tracked vehicles, as well as certain other features, were not yet functional.[22] The JSTARS aircraft can form detailed images of the ground (in so-called synthetic-aperture radar mode) while simultaneously scanning for mobile targets in a search mode.[23]

Phased-array technology is also used for radio eavesdropping. A single 100-meter-wide antenna on the Trumpet satellite, for example, can monitor thousands of radio or telephone conversations at once from an elliptical orbit ranging from a few hundred miles to 23,000 miles above earth.[24]

A number of other improvements in radar technology are afoot. Miniaturization of electronic components is improving the performance of radars on aircraft and satellites, for example. This promises major improvements. To give a sense of the rate of improvement, JSTARS had resolutions of only about twelve feet during the Persian Gulf War, but radars small enough to fit on unmanned aerial vehicles (UAVs) may soon have resolutions better than a foot.[25]

To counter the predictability of satellites' orbits that allow other countries to hide some activities from current KH-11 and Lacrosse systems, and to provide more continuous information to warfighters, constellations of radar satellites may be built to provide good-resolution images within minutes.[26] At present, the so-called Discoverer-2 program is in development; it would deploy some two dozen satellites into low-earth orbit for this pur-

21. Tony Perry, "High-Tech Spy System Could Transform War," *Los Angeles Times*, March 17, 1997, p. A1.

22. Northrop Grumman Corporation, "Joint STARS Data Analysis of 'The Battle of Khafji,' Final Report" (Melbourne, Fla.: Northrop Grumman Corporation's Surveillance and Battle Management Systems, May 1997), pp. 7–19.

23. David A. Fulghum, "Upgrades to Sharply Improve Radar's Vision," *Aviation Week and Space Technology*, June 22, 1998, pp. 57–58.

24. Craig Covault, "NRO Radar, Sigint Launches Readied," *Aviation Week and Space Technology*, September 1, 1997, p. 22.

25. "Gnat Exploits Advanced SAR," *Aviation Week and Space Technology*, August 30, 1999, p. 30; David Mulholland, "New U.S. Radar Increases Possibilities for UAVs," *Defense News*, September 20, 1999, p. 20.

26. Warren Ferster, "U.S. Wants Radar Satellite," *Defense News*, March 16–22, 1998, p. 4; Walter Pincus, "2 New Tests Again Catch U.S. Intelligence off Guard," *Washington Post*, May 14, 1998, p. A28.

pose, though the current projected cost per satellite of $100 million (in contrast to the Lacrosse radar satellite's reported cost of some $1 billion) casts some doubt on the realism of expectations about this program.[27] Radar-satellite constellations could also be designed as clusters of small satellites, maintaining separations of enough distance to prevent all from being disabled by a single localized explosion or "hit-to-kill" antisatellite weapon, but being close enough to work together as a synthetic-aperture radar.[28]

Among the most important areas of radar and radio research are attempts to design radars that can penetrate below the forest canopy (though probably not through large branches or tree trunks), picking up signals from buildings or roads. Certain lower-frequency, longer-wavelength radars do penetrate foliage (and, to a degree, certain types of soil). Work is proceeding on these systems and on the automatic target recognition software needed to make them practical.[29] Other new radars may be able to locate shallowly buried objects like land mines and some bunkers down to depths of several meters.

It may even be possible to "see through walls" with nearby microwave radars involving very wide frequency bands. In lower-power versions, these types of devices may also prove useful for short-range communications that would be extremely difficult for an enemy to detect, because only very limited amounts of energy would be transmitted at any given frequency.[30]

Radars may get somewhat more powerful in the future, too, though perhaps not radically so. For example, use of gallium arsenide semiconductors may allow a severalfold increase in power and perhaps a doubling in a given radar's detection or tracking range.[31]

Bistatic radars, in which a transmitter is located in one place and the receiver elsewhere, may make a comeback—this time as a partial antidote to stealth technology. Bistatic radars were once popular because they helped keep noise levels low in the receiver, which was located hundreds of meters,

27. Robert Wall, "Space-Based Radar Development Begins," *Aviation Week and Space Technology*, March 1, 1999, p. 33; Michael O'Hanlon, *U.S. Costs of Verification and Compliance under Pending Arms Treaties* (Congressional Budget Office, September 1990), p. 46.

28. Warren Ferster, "U.S. Air Force, NASA Eye Small Satellite Operations," *Defense News*, August 17–23, 1998, p. 10.

29. John G. Roos, "Unmasking the Enemy," *Armed Forces Journal International*, vol. 135 (April 1998), p. 51; Beth Jannery, "DARPA Taps Lockheed Martin for FOPEN SAR," *Journal of Electronic Defense* (November 1997), pp. 30–31.

30. William B. Scott, "Radar Patent Skirmish Ripples through Industry," *Aviation Week and Space Technology*, June 15, 1998, pp. 26–27.

31. Director, Defense Research and Engineering, *Joint Warfighting Science and Technology Plan*, p. IV-D-11.

or even kilometers, from the transmitter. In the case of stealth, bistatic radars would be desirable because many stealth surfaces work by redirecting radar energy to make sure it does not return along the path from which it came. So if one is in the right position to listen for an echo, it may still be picked up. The signal may be relatively diffuse, but it can be considerably stronger than along the axis of origin.[32] A form of passive bistatic radar may also hold promise—one that relies on the increasingly dense background of civilian radio and television broadcasts present virtually everywhere on earth. When an object interrupts the normal background pattern of such transmissions at a receiver site, it may be detectable at militarily significant distances.[33]

Stealth will also be challenged by radars operating at different frequencies. To a large extent, an object's radar reflectivity depends on the relationship between its dimensions and the wavelength of the incoming radar. By changing the latter, an observer may make an object stand out much better. For example, higher-frequency millimeter-wave radar may be used to counter stealth to some extent. Although it is not as effective as lower-frequency radars in seeing through water or even rain, it could be very useful in a number of short- and medium-range settings. Coupling such radars with good surface-to-air missiles could improve future air defenses against stealthy aircraft.[34]

Given the ongoing trends in miniaturization, it may eventually prove possible to place affordable precision radars inside small precision-guided munitions. This would represent a major breakthrough by allowing such weapons to track and attack mobile targets autonomously even during bad weather. (Radar transmitters and receivers are already deployed on air-to-air missiles, which are larger and more expensive weapons and which do not need to cope with a background as complex as an air-to-ground munition would face when approaching its target.)

As for radio transmissions, increasingly large and sensitive antennas in space or on aircraft will make it possible to monitor faint emissions from weapons or other sensitive technologies that have been tagged with small electronic devices. Similarly, cell telephone technology is likely to evolve to the point where wrist telephones can have their signals amplified and relayed by satellites, allowing communication between individuals at any point on earth.[35]

32. Fulghum, "Upgrades to Sharply Improve Radar's Vision," pp. 58–59.

33. Bryan Bender, "Surveillance System Uses Broadcast Signals," *Jane's Defence Weekly*, October 21, 1998, p. 10.

34. Robert Wall, "U.S. Helicopters Face More Capable Missiles," *Aviation Week and Space Technology*, January 18, 1999, p. 25; David A. Fulghum, "New Radars Peel Veil from Hidden Targets," *Aviation Week and Space Technology*, January 18, 1999, pp. 58–60.

35. Kenneth L. Adelman and Norman R. Augustine, *The Defense Revolution: Strategy for the Brave New World* (San Francisco: Institute for Contemporary Studies, 1990), p. 71.

Other Sensors

Detection techniques in this broad category include seismic monitoring of nuclear tests, sonar detection and tracking of underwater objects, short-range motion and magnetic detectors, particle beams, chemical and biological sensors, and acoustic devices.

Seismic monitoring can pick up signals from underground nuclear tests somewhat less powerful than the Hiroshima and Nagasaki bombs (roughly ten to twenty kilotons). It can estimate their power within perhaps a factor of two to three uncertainty.[36]

Sonar can be both passive and active. A sonar sensor can either emit sound waves and then listen for their echo or simply listen for the emissions of a ship or submarine. Depending on water conditions, proximity to shore, and the sound generated by a target vessel, sonar can permit detections at various ranges from several to a few hundred kilometers. For example, active sonar might work out to ranges of about twenty to fifty kilometers, depending largely on water conditions. Passive sonar varies much more as a result not only of water conditions but also of the noise emitted by the target; ranges can vary from just a few kilometers against a modern diesel-electric submarine operating on battery power or an advanced nuclear submarine to tens or hundreds of kilometers against an older sub running diesel engines or moving at high speeds. Consider one example of what these detection ranges mean in practice: Assume that several dozen sonar-carrying platforms were devoted to an active search effort for a submarine known to be in a body of water like the Sea of Japan or the Persian Gulf (an area of around 100,000 square miles). In that event, a quiet submarine in such waters would probably be found within a day or so.[37]

A number of motion and acoustic sensors are useful in tactical settings for detecting nearby activity such as helicopter flights, gunfire, and human

36. Steve Fetter, *Toward a Comprehensive Test Ban* (Cambridge, Mass.: Ballinger, 1988), pp. 107–58.

37. James F. Dunnigan, *How to Make War*, 3d ed. (William Morrow, 1993), pp. 251–55; Owen Cote and Harvey Sapolsky, *Antisubmarine Warfare after the Cold War* (Security Studies Program, Massachusetts Institute of Technology, 1997), pp. 11–13; Rear Admiral W. J. Holland, "Battling Battery Boats," *Proceedings* (June 1997), p. 31; Tom Stefanick, *Strategic Antisubmarine Warfare and Naval Strategy* (Lexington, Mass.: Lexington Books, 1987), pp. 35–37, 49. Stefanick estimates that roughly forty U.S. attack submarines would have required fourteen hours to search the Sea of Okhotsk for Soviet submarines, under ideal sonar conditions. Assuming that the number of sonar platforms doubled, and the amount of sea was reduced by a factor of five, would make the job ten times easier. But detection ranges could, as Stefanick notes, easily be ten times worse than the thirty nautical miles he assumes, particularly in shallow water, canceling out the other effects.

foot traffic.[38] Such sensors were used in Vietnam. But they lacked capabilities like global positioning system (GPS) locators and on-board microprocessors to allow their positions to be accurately known and to identify the probable source of any sound (or magnetic or seismic signals). Future acoustic and motion sensors will have more advanced capabilities. They will be able to locate, and quite often identify, a number of types of nearby targets.[39] Costs are declining enough to put such sensors on $50,000 wide-area munitions that could be used, among other things, as alternatives to anti-armor land mines.[40] Costs are likely to decline further, perhaps even by as much as a factor of ten or so.[41]

Chemical weapons detection and agent identification sensors function in two main ways: they either bombard gases with light to observe which frequencies are absorbed or measure the degree to which molecules are adsorbed on inert surfaces. The first technique is spectroscopy; the second, chromatography. The first technique is used in most sensors today but is prone to false alarms and inaccuracies.

Improvements in robotics will be of considerable use for chemical and biological weapons detection. Complete laboratories may be deployable on mini-UAVs or small wheeled vehicles resembling toys; alternatively, robotics may simply gather samples and return them to a larger facility for processing. Most of these microrobotics will have limited ranges, but they will nonetheless be very valuable in certain settings, such as urban environments—particularly when one already has a general sense of where to search for agents. Plans are on the drawing board to develop complete chemical weapons detector systems roughly the size of a shoe box.[42]

There is some hope of pursuing the spectroscopy concept for biological weapons detection. However, near-term goals are much more modest: sim-

38. Mark Hewish, "Silent Sentinels Lie in Wait: Unattended Ground Sensors," *Jane's International Defense Review,* vol. 31 (January 1998), pp. 48–52.

39. Vick and others, *Enhancing Air Power's Contribution against Light Infantry Targets,* pp. 26–27; Libicki, "DBK and Its Consequences," pp. 26–27; Hewish, "Silent Sentinels Lie in Wait," pp. 48–52.

40. Robert Braham, "Technology 1999 Analysis and Forecast: Aerospace and Military," *IEEE Spectrum,* vol. 36 (January 1999), p. 76.

41. Charles C. Carson and Steven G. Peglow, "A White Paper on the Potential for Distributed Ground Sensors in Support of Small Unit Operations," in Defense Science Board 1996 Summer Study Task Force, *Tactics and Technology for 21st Century Military Superiority,* vol. 3, *Technology White Papers* (Department of Defense, 1996), p. V-32.

42. David Mulholland, "Shrinking Chemical-Weapon Sensors Speed Analysis," *Defense News,* August 3–9, 1998, p. 10.

ply to be able to compress reagent-based (that is, "wet") biological weapons detectors into man-portable packages.[43]

The use of miniaturized sonar sensors, be they on midget submarines or unmanned underwater vehicles deploying from a mother sub or simpler devices, will increase as well. They will also work increasingly as networks connected via fiber-optic cable, extending the search footprint of any given submarine to shallower waters and reducing the vulnerability of humans in the mother submarines.[44]

Acoustic sensors will also help to detect sniper fire. Specifically, they will sense the shock waves created by the (generally supersonic) bullets. In a manner similar to the way sonar determines the bearing to a target, they will be able to determine the direction from which the shot came with at least fair accuracy.[45] Acoustic sensors are also becoming quite important for other purposes. The army's brilliant antitank weapon (BAT) is a submunition that will use listening devices to detect—and even discriminate—enemy tanks before it attacks them. Finally, acoustic devices may be used to characterize an agent inside a container. Immediate access to the container at issue is needed, but it is becoming possible to identify certain types of chemical agents by the peaks and valleys in their sound-frequency measurement patterns.[46]

Another category of sensors involves the physics of high-energy particles and radiation. Passive detectors register the emissions of a material like uranium. Active devices generate their own radiation, be it particles like neutrons or short-wavelength (high-energy) electromagnetic radiation like X-rays, and wait for any return signal.

43. Vernon Loeb and Walter Pincus, "New Device Can Sense Germ Arms," *Washington Post*, March 3, 1999, p. A20; Robin Lloyd, "Lab on a Chip May Turn Police into DNA Detectives," *Washington Post*, March 1, 1999, p. A9.

44. Robert Holzer, "Russian Subs' Stealth Prompts U.S. Response," *Defense News*, March 20–26, 1995, p. 3; Robert Holzer, "U.S. Navy Manta May Expand Sub Combat Power," *Defense News*, March 17–23, 1997, p. 1; David Foxwell, "Sonars Go Shallow," *Jane's International Defense Review*, vol. 30 (October 1997), p. 48; Office of Naval Intelligence, *Worldwide Submarine Proliferation in the Coming Decade* (1995); Robert Holzer, "Dangerous Waters," *Defense News*, May 4–10, 1998, p. 1; General Accounting Office, *Navy Mine Warfare*, GAO/NSIAD-96-104 (March 1996), pp. 22–23; Seffers, "Special Operations Forces Want to Deploy with Robots," p. 3; Director, Defense Research and Engineering, *Joint Warfighting Science and Technology Plan*, pp. IV-G-14 through IV-G-15; Buzz Broughton and Jay Burdon, "The (R)evolution of Mine Countermeasures," *Proceedings*, vol. 124 (May 1998), p. 57.

45. Hewish and Pengelley, "Snipers Fire at Their Peril," pp. 30–36.

46. William B. Scott, "Noninvasive Detector Identifies Chemical Weapon Agents," *Aviation Week and Space Technology*, May 18, 1998, pp. 69–70; McKenna, "Combating Chemical/Biological Attacks."

The Realities of Physics—and of Countermeasures

For all of the impressive current and projected future attributes of these sensors, two main principles constrain their long-term potential. The first is, in essence, a law of war: for every new sensor developed, countervailing efforts will be made to defeat it. The second is that the laws of physics limit what certain types of sensors are inherently capable of accomplishing.[47]

VISIBLE AND NEAR-VISIBLE LIGHT SENSORS. To begin, as the war in Kosovo has reminded those who had forgotten, optical and infrared sensors cannot see through bad weather. The same thing was, of course, true in Desert Storm.[48] It will also remain true in the future because of basic laws of physics. Water absorbs virtually all radiation wavelengths in this part of the electromagnetic spectrum. Filters and selective frequency bands can be used to minimize the effects of haze and smoke but not to penetrate heavy clouds or rain.[49] (This means, moreover, that optical lasers will remain unsuitable for antisubmarine warfare or mine warfare for water depths greater than about 100 meters.[50])

It is also worth emphasizing the obvious: that optical and infrared sensors cannot see inside most solid objects. The implications of this fact are that weapons contained inside cars and trucks, or hidden within buildings or behind trees, will not be identifiable through these types of sensors— unless the behavior of those hiding the weapons somehow reveals hostile

47. There has been little writing on this subject to date in the policy literature. For two important exceptions on the subject of satellites, see Paul B. Stares, *Space and National Security* (Brookings, 1987), p. 40; Ashton B. Carter, "Satellites and Anti-Satellites: The Limits of the Possible," *International Security*, vol. 10 (Spring 1986), pp. 46–98. A key point made in these studies is that high-resolution sensors on low-orbit satellites, individually or in small numbers, cannot provide anything close to continuous coverage of a given point on earth. Because of the curvature of the earth, their view is blocked beyond a certain point. As a result, it would take at least thirty satellites to maintain continuous coverage at 300 miles' satellite altitude, for example.

48. See Michael O'Hanlon, "Should Serbia Be Scared?" *New York Times*, March 23, 1999, p. A31; Thomas A. Keaney and Eliot A. Cohen, *Gulf War Air Power Survey Summary Report* (Government Printing Office, 1993), pp. 16, 171–73.

49. David A. Fulghum, "Darkstar Camera Takes First Images," *Aviation Week and Space Technology*, May 5, 1997, p. 29.

50. Statement of Rear Admiral Richard D. Williams III before the Senate Armed Services Subcommittee on Expeditionary Warfare, March 19, 1996, p. 8; Broughton and Burdon, "The (R)evolution of Mine Countermeasures," p. 57; Richard L. Garwin, "Will Strategic Submarines Be Vulnerable?" in Steven E. Miller and Stephen Van Evera, eds., *Naval Strategy and National Security* (Princeton University Press, 1988), p. 235.

Water transmits most other kinds of electromagnetic radiation even less. Only long-wavelength/low-frequency radio waves, in addition to light, can penetrate seawater as far as a few meters. John David Jackson, *Classical Electrodynamics*, 2d ed. (John Wiley and Sons, 1975), pp. 290–92. See also Carter, "Satellites and Anti-Satellites," p. 69.

intent. Military vehicles designed to appear like civilian vehicles on the outside also will be hard to identify with optical and infrared sensors. Decoys such as fake tanks, or older tanks that may or may not still be functional, may be hard to distinguish from the real thing from distances of several kilometers.[51]

Multispectral and imaging infrared sensors will help discriminate real targets from decoys in some cases. For example, they could reveal heat distributions on a vehicle that may provide evidence of a large, hot tank engine lurking below. However, decoys, flares, chaff, smoke, and other countermeasures have always gotten better as sensors have improved, and the trend will surely continue. For example, today some aircraft decoys are made with highly aerodynamic shapes so they will not slow down quickly once released; that makes it harder for heat-seeking missiles to recognize them by their trajectories. Tunable lasers on infrared decoys can cover a wide frequency spectrum, presenting false targets to infrared sensors across the whole spectrum.[52] Tank decoys could be built with simple heat sources that create infrared signatures indicative of a hot engine. Weapons like armored personnel carriers could be built with external skeletons and skins designed to resemble civilian vehicles.[53]

RADAR AND RADIO EMISSION SENSORS. Radar and radio waves can penetrate clouds and rain (particularly at lower frequencies) and, to a lesser extent, foliage and soil. But they cannot physically penetrate deeply into soil or water. Nor can they reach inside sealed metal containers. That is due to a basic law of physics: any electromagnetic radiation is stopped within fractions of a centimeter at the surface of a conductor.[54] This means that mortars and missiles inside well-sealed truck bays, weapons built into commercial airliners, and missile launchers on commercial ships will be difficult, if not impossible, to find by radar, optical, and near-optical sensors. This is true not just in the near term but indefinitely.

To some extent these facts can be viewed as natural limitations on the effectiveness of radar. They also open up possibilities for an adversary to develop countermeasures that exploit those limitations. For example, as sug-

51. Steven Lee Myers, "Damage to Serb Military Less than Expected," *New York Times*, June 28, 1999, p. A1.

52. See Bruce D. Nordwall, "Laser Countermeasures Developed for Tactical Aircraft," *Aviation Week and Space Technology*, October 27, 1997, pp. 52–53; David A. Fulghum, "Enemies Face Frustration as Aerial Decoys Improve," *Aviation Week and Space Technology*, October 27, 1997, pp. 67–69.

53. Libicki, "DBK and Its Consequences," pp. 44–48; Vick and others, *Enhancing Air Power's Contribution against Light Infantry Targets*, p. 58.

54. Jackson, *Classical Electrodynamics*, pp. 23, 298, 337.

gested previously, tanks and other armored vehicles could be designed with shapes and external skins that resemble commercial vehicles, making it very difficult to identify them from a distance.[55] Doing so might not be practical in all settings. But tanks and other armored vehicles simply moving about a theater of operations might be equipped with snap-on metal skins that would make their radar signature from overhead identical to that of a commercial truck. Acoustic sensors might still be able to identify the tank, but they are generally of much shorter range and likely to remain so (as discussed further below).

Another fundamental constraint on radar is that it is an active, or energy-emitting, detection technology. That means that an object emitting radar gives away its own location even as it searches for enemy targets—opening up many possibilities for the successful use of countermeasures by the adversary. The undesirability of radiating strongly is underscored by the fact that modern stealth aircraft are designed to minimize their own use of radar. The F-117 Raven stealth fighter does not even have a search and targeting radar. The B-2 bomber has one, but it is designed to emit at the lowest practical power and have the narrowest possible beam in order to make tracking difficult.[56] As such, it is better for confirming a target's location on final approach than for conducting a full-fledged search. Highly directional wide-bandwidth radar beams minimize the chances of detection and tracking but do not eliminate them—particularly against future listening networks that will themselves be more sophisticated. The radar signal will always be much stronger at the site of the target being illuminated than back at the receiving antenna; the strength of any signal weakens as the square of distance from its source. Because of this, a target that possesses a radar receiver of its own will generally know it is being examined before weapons can be launched at it. As a result, it will often have the opportunity to take defensive action: shooting at the enemy, turning on an array of radar jammers that emit much stronger waveforms than the echo from the target itself, or simply ducking for cover.[57]

55. Libicki, "DBK and Its Consequences," pp. 44–45.

56. Colonel Timothy M. Laur and Steven L. Llanso, *Encyclopedia of Modern U.S. Military Weapons* (New York: Berkley Books, 1995), pp. 80–81; David Mosher, *Options for Enhancing the Bomber Force* (Congressional Budget Office, July 1995), p. 87.

57. See Skolnik, *Introduction to Radar Systems*, pp. 30–61; J. C. Toomay, *Radar Principles for the Non-Specialist* (Belmont, Calif.: Lifetime Learning Publications, 1982), pp. 7, 74–88; Bill Sweetman, "The Progress of the F-22 Fighter Program," *Jane's International Defense Review*, vol. 30 (March 1997), p. 17; Ben R. Rich and Leo Janos, *Skunk Works* (Boston: Little, Brown, 1994), pp. 26–27, 36–38, 82; Philip J. Klass, "Towed Decoys, Flares Displayed," *Aviation Week and Space Technology*, January 4, 1999, p. 46; and Lane Pierrot, *A Look at Tomorrow's Tactical Air Forces* (Congressional Budget Office, January 1997), p. 77.

One possible result is that large aircraft like the airborne warning and control system (AWACS) and JSTARS, which radiate heavily and present large targets to antiair missiles, may prove more vulnerable in the future. They will not be able to fly simply farther and farther away from the battlefield indefinitely in order to stay out of range of antiair missiles. Beyond a certain distance, the curvature of the earth will prevent them from seeing what is on the ground or in the lower atmosphere. For example, at 30,000 feet altitude—around the operating altitude of today's JSTARS—an aircraft's radar cannot see ground objects beyond 300 miles (and ranges are less if targets are located in valleys or on the back sides of hills and mountains).[58] If the surveillance aircraft flies near enough front lines of battle to try to see deeply into enemy territory, it may prove vulnerable to air-to-air or surface-to-air missiles.

Active sensors like radar also require large amounts of power. This is particularly true for satellites operating many hundreds of kilometers away from objects on the ground that they are trying to detect or track. The return echo from a radar falls off in strength as the fourth power of distance, meaning that doubling the distance between radar and target makes the echo sixteen times weaker.[59] Wide-area continuous coverage of targets from space-based radar constellations may prove possible by 2020—at least to a certain resolution—but such accomplishments cannot yet be assumed.[60]

There are potential missions for radar that, while theoretically possible, are proving very difficult to accomplish. One is using radar to track the wakes of submarines, a problem that requires data integration over considerable periods of time to filter out the effects of natural waves and other surface disturbances to reveal the underlying signal of a submerged vessel.[61] Moreover, even if this technology someday proves workable, it will be hard to use against submarines that are frequently changing direction and depth, since their wakes will be constantly changing and thus hard to separate out of a background noise pattern.[62] The problem is so daunting that a recent

58. See Michael E. O'Hanlon, *The Art of War in the Age of Peace* (Westport, Conn.: Praeger, 1992), p. 120; David A. Fulghum, "JSTARS Battles SAMs and Bias at Green Flag," *Aviation Week and Space Technology*, June 22, 1998, p. 56.

59. M. Thomas Davis, "Sensor's Acquisition Strategy Needs Revisiting by Pentagon," *National Defense*, vol. 83 (December 1998), p. 31.

60. Michael Gruber, "UAV-Based Sensing for Surveillance and Targeting," in Defense Science Board 1996, *Tactics and Technology for 21st Century Military Superiority*, vol. 3, p. II-8.

61. Stares, *Space and National Security*, pp. 42–43.

62. William J. Broad, "U.S. Loses Hold on Submarine-Exposing Radar Technique," *New York Times*, May 11, 1999, p. A6.

Defense Science Board task force stated flatly that "submarine operations also enjoy inherent protection . . . from satellite surveillance."[63]

Foliage-penetrating radar appears to be a more promising technology. However, it suffers from three significant challenges itself. First, it must use relatively long-wavelength radar to penetrate foliage. Wavelengths are generally around one meter in length, meaning that resolution of the features of a target object is generally limited to about that same one-meter clarity. So the radar may not do a good job of distinguishing a military vehicle from a truck, even if it can detect them both. This problem is serious. Indeed, even high-resolution radars on aircraft such as the F-15E had difficulty distinguishing tanks from trucks at tactical distances during Operation Desert Storm.[64] It will be especially difficult to overcome this problem when using longer wavelengths, as required in foliage-penetrating radars. Second, the prevalence of strong radar returns from objects such as tree trunks provides a great deal of clutter that must be distinguished from real targets. Again, this poses a serious challenge, particularly with radar of only moderate resolving power. In practice, this means that foliage-penetrating radars must use various frequency bands, examine objects from various angles, and use formidable computing power and software to integrate information in the effort to find real targets.[65] The third constraint does not concern foliage-penetrating radar per se, but rather the problem of attacking forces within forests. Even if military objects or formations are identified, it can be hard to attack them. Tree branches and trunks would provide people and vehicles protection against incoming rounds.[66]

A constraint on radio-wave-based sensors in general is that they usually cannot detect radiation that is not emitted into the atmosphere. Again, this is a natural constraint upon radar that opens up many possibilities for the use of countermeasures by an adversary. For example, the increasing prevalence of fiber-optic cable will make it possible for many countries to reduce their dependence on radio, at least for communications such as contacts between a national government and its ground-based commanders in the

63. Defense Science Board Task Force, *Submarine of the Future*, briefing charts, Office of the Under Secretary of Defense for Acquisition and Technology, 1998.

64. See General Accounting Office, *Operation Desert Storm: Evaluation of the Air Campaign*, GAO/NSIAD-97-134 (June 1997), pp. 32–36.

65. Lam Nguyen, Ravinder Kapoor, and Jeffrey Sichina, "Detection Algorithms for Ultra-Wideband Foliage Penetration Radar," *Proceedings of the International Society for Optical Engineering* (April 1997), pp. 165–77.

66. Vick and others, *Enhancing Air Power's Contribution against Light Infantry Targets*, pp. 17–20; Karl A. Kappra and others, "Ultra-Wideband Foliage and Ground-Penetrating Radar Experiments," U.S. Army Research Laboratory, Adelphi, Md., 1999.

field. This will seriously constrain the ability of the United States to eaves-drop. Communications that were monitored in the past will, in many cases, not be accessible to an outside listener unless the listener can tap into land lines carrying the fiber-optic cable—always a difficult proposition.[67] Enemy forces that are maneuvering, flying, or sailing about may be hard-pressed to disguise their communications, giving the United States important advantages in head-to-head or symmetric warfare. The advantages will be less notable, however, in other combat settings. In addition, improving encryption technologies may make it harder for the United States to decipher any signals it does obtain—though the verdict on encryption is less clear-cut.[68]

OTHER SENSORS. Acoustic, magnetic, and motion sensors generally suffer from severe and innate range constraints. Today, maximum ranges of acoustic detectors are typically a few kilometers against small numbers of people and several tens of kilometers against vehicles or aircraft.[69] Ranges can commonly be less due to wind and other natural noise.

Ranges are typically tens of meters for localized seismic ground sensors.[70] Magnetic fields fall off quickly with distance from their source—to be precise, as the third power of distance, meaning that an observer at ten meters from a source will experience a field 1,000 times weaker than an observer at one-meter distance.[71] As a practical matter, this means that vehicle-size objects are magnetically detectable out to no more than hundreds of meters and, more commonly, to just tens of meters (weapon-bearing individuals are detectable to distances of only a few meters).[72]

The sensitivity of seismic sensors for detecting nuclear tests is also unlikely to improve greatly. At present, sensors are not adequate to detect tests of yields smaller than one kiloton, as was recently underscored during the testing of Indian and Pakistani weapons (only three were apparently picked up by outside intelligence; others were not). Seismic sensors can usually detect

67. Jeffrey T. Richelson, *The U.S. Intelligence Community*, 2d ed. (Cambridge, Mass.: Ballinger, 1989), p. 170.

68. Libicki, "DBK and Its Consequences," p. 26.

69. David A. Fuess, "Unattended Ground Sensor Technology," in Alex Gliksman, ed., *Meeting the Challenge of International Peace Operations: Assessing the Contribution of Technology* (Livermore, Calif.: Center for Global Security Research, Lawrence Livermore National Laboratory, 1998), pp. 136–37.

70. Robert A. Zirkle and others, *Advanced Sensor and Information Processing Technologies for Peacekeeping Operations*, IDA Paper P-3112 (Alexandria, Va.: Institute for Defense Analyses, 1996), p. III-16.

71. Jackson, *Classical Electrodynamics*, p. 178.

72. Zirkle and others, *Advanced Sensor and Information Processing Technologies for Peacekeeping Operations*, p. E-5.

and identify tests more powerful than one kiloton, though they may have trouble picking up signals from detonations of less than ten kilotons if they are carried out in large underground cavities. Even if the additional ground stations called for in the comprehensive test ban treaty are eventually built, the situation will not change greatly in many areas.[73]

There are fundamental limitations on nuclear material detectors as well. For passive detectors, a nuclear warhead's normal signature quickly becomes weak against the earth's natural radioactive emissions; beyond 100 meters, it will be smaller than the normal fluctuations in the natural radioactive background and thus difficult—if not impossible—to detect.[74] Some improvement is possible through use of superconducting spectrometers and other innovations, but the phenomenology will remain short range.[75]

Active sensors may do a bit better, but they will also be constrained. For example, only about one-ten-thousandth of a neutron beam's initial strength will survive a transit of one kilometer through the atmosphere. This effect, and similar ones for other potential active interrogators, make the prospects for long-range detection of even unprotected and unshielded nuclear materials on the earth's surface very poor—unless they have physically been brought into a verification regime by becoming electronically tagged.[76]

Biological detectors based on laser techniques will not allow detection of agents that have not yet been released into the atmosphere. Nor, for the foreseeable future, will they allow positive identification of specific agents that been released. They may be able to identify certain amino acids, but those occur in all life.

73. Larry S. Walker, *A Systems Perspective of Comprehensive Test Ban Treaty Monitoring and Verification*, SAND 96-2740 (Albuquerque, N.M.: Sandia National Laboratories, 1996); Fetter, *Toward a Comprehensive Test Ban*, pp. 114–15, 138–40.

74. See Roald Z. Sagdeev, Oleg F. Prilutsky, and Valery A. Frolov, "Passive Detection of Nuclear-Armed SLCMs," in Frank von Hippel and Roald Z. Sagdeev, eds., *Reversing the Arms Race: How to Achieve and Verify Deep Reductions in the Nuclear Arsenals* (New York: Gordon and Breach, 1990), p. 372; S. T. Belyaev and others, "The Use of Helicopter-Borne Neutron Detectors to Detect Nuclear Warheads in the USSR-US Black Sea Experiment," in von Hippel and Sagdeev, *Reversing the Arms Race*, p. 399.

75. Simon E. Labov, "Superconducting, Superresolving Gamma-Ray Spectrometers," in *Arms Control and Nonproliferation Technologies: Radiation Detection* (Livermore, Calif.: Lawrence Livermore National Laboratories, 1997), pp. 14–15.

76. Samuel Glasstone, ed., *The Effects of Nuclear Weapons*, rev. ed. (Government Printing Office, 1962), pp. 387–94; Steve Fetter and others, "Detecting Nuclear Warheads," in von Hippel and Sagdeev, *Reversing the Arms Race*, p. 292.

Communications and Computers

Once sensors have accumulated data, the information must be brought together, processed, and disseminated to those in the field in a position to make good use of it. These areas of defense operations probably offer the greatest potential for huge strides in coming years as a result of technological progress. Even so, there are important limits on what can be expected.

Communications

There are several major technological reasons why communications technologies can be expected to keep improving rapidly over the next two decades. They range from the miniaturization of electronics—making possible smaller radars and radios as well as easily affordable identification-friend-or-foe (IFF) devices and the like—to a proliferation of satellites to technologies such as laser communications systems and fiber-optic cables.

Today's information and communications systems are largely service-specific and analog. In the years ahead they will be integrated between military services. They will also be digitized.[77] Digital radio can transmit impressive amounts of data among a large group of weapons platforms or soldiers.[78]

Digitized information networking on the battlefield is just beginning to come into its own, and major progress will occur as various systems are put in place. For example, experiments have recently been conducted to transfer targeting data originating at an imaging satellite via an AWACS aircraft to an F-15E Strike Eagle fighter over the joint tactical information distribution system (JTIDS).[79] This type of rapid routing of information from a satellite to a weapons platform has generally not been possible to date.[80] In NATO's 1999 war against Serbia, new software and procedures were used

77. Jeffrey R. Cooper, *The Emerging Infosphere* (McLean, Va.: Science Applications International Corporation, 1997), pp. 32–35; Kenneth Allard, "An Agenda for Fighting Cyberwars," *Los Angeles Times*, March 31, 1997, p. A11.

78. For example, a variant of the Joint Tactical Information Distribution System (JTIDS) divides each second up into more than 100 time slots; within each of those time slots at least fifty bits of data can be exchanged, making for a total of several thousand bits of data per second. Other versions can transmit up to 250,000 bits per second. Kenneth Allard, *Command, Control, and the Common Defense*, rev. ed. (National Defense University, 1996), pp. 215–16; Mark Hewish and Joris Janssen Lok, "Connecting Flights: Datalinks Essential for Air Operations," *Jane's International Defense Review*, vol. 136 (December 1998), p. 42.

79. William B. Scott, "'Strike' to Include U-2, AWACS, Joint-STARS," *Aviation Week and Space Technology*, March 11, 1996, p. 56.

80. David A. Fulghum, "New Focus: Battle Command," *Aviation Week and Space Technology*, June 22, 1998, p. 60.

to hasten the transfer of satellite reconnaissance data and other tactical intelligence to warfighters. In some cases they received it within minutes, rather
than the hours or days that had often been necessary before.[81] Navy ships
in a given battle group can now exchange data between them at the rate of
128 kilobits per second or more.[82] Progress in computers means that most
sensor devices and platforms can now also process some of the information
they obtain. Only the key results, rather than the vastly larger raw data
streams from the radar returns themselves, need be broadcast by radio.[83]

The army continues to equip its troops with older radios that require
line-of-sight access to hilltop transmitters to work properly.[84] This antiquated
approach should change soon, however, simply by making good use of technology already available. The army has recently equipped the 900 vehicles
in its experimental force, the 1st Brigade of the 4th Infantry Division, with
appliqués—that is, a GPS locator, digital radio, and computer. Costs are
quite modest.[85] Altogether, the hardware costs appear to have totaled about
$250 million[86]—a modest amount when compared with the other equip-
ment costs for a heavy brigade of well over $1 billion (moreover, the 1st
Brigade's appliqué equipment was largely tailor-made).[87]

The army's "tactical internet" can also be tied in with UAVs and JSTARS
aircraft. It shares information on friendly locations as well as on enemy
forces. (It also retains voice communication capabilities, both as a backup
and for the close-in battle, where computer screens proved unwieldy.) In
early tests these communications were not jammed or otherwise challenged;

81. Warren Ferster, "Troops Receive Target Data from Satellite Images," *Defense News*,
July 26, 1999, p. 42; Nick Cook, "War of Extremes," *Jane's Defence Weekly*, July 7, 1999,
p. 21; David A. Fulghum, "Data Link, EW Problems Pinpointed by Pentagon," *Aviation Week
and Space Technology*, September 6, 1999, pp. 87–88.

82. Robert Holzer, "U.S. Navy: IT-21 Shines in Debut Desert Fox Action," *Defense News*,
January 25, 1999, p. 6.

83. Allard, *Command, Control, and the Common Defense*, pp. 209–15; Toomay, *Radar
Principles for the Non-Specialist*, p. 125.

84. Pat Cooper, "Bosnia Study Highlights U.S. Communication Inadequacies," *Defense
News*, January 13, 1997, p. 14.

85. Reportedly a "reinforced" commercial computer costing $20,000 performed better
than a $100,000 military-specifications model. See Mark Hanna, "Task Force XXI: The Army's
Digital Experiment," *Strategic Forum*, no. 119 (July 1997), pp. 1–4. Radios making up the
enhanced position location reporting system each cost $60,000 to date and can be expected
to cost just half that in the future. See George I. Seffers, "U.S. Army Considers Boosting EPLRS
Radio Purchase," *Defense News*, September 15, 1997, p. 23.

86. Steven Komarow, "Cybersoldiers Test Weapons of High-Tech War," *USA Today*,
March 6, 1997, p. A1.

87. Frances Lussier, *Budgetary and Military Effects of a Treaty Limiting Conventional Forces
in Europe* (Congressional Budget Office, 1990), p. 34.

future versions will have to be hardened and made more redundant to be proven dependable in battle.[88] Nor were command posts put under simulated attack and forced to move in these tests; at present they are far too unwieldy to relocate quickly. The army's tactical internet also involves several separate networks with limited interoperability between them at this point.[89] All of these kinks should be worked out with time.

Data compression techniques are allowing more information to be transmitted through the electromagnetic spectrum as well. Raw digital video containing some 45 million bits per second of information can now be transmitted in a data stream of 1.2 million bits per second. This effect is achieved by eliminating or reducing the transmission of redundancies and patterns in the data.[90] For example, if a certain part of an image does not change between frames, the absence of additional information about that frame might be correctly understood by the software reconstructing the image on the receiving end as a steady image. Using similar data compression techniques, a recognizable image of a human face can now be stored with one-two-hundredth the number of bits previously required.[91]

Laser and fiber-optic communications are remarkable new innovations as well. However, they require line-of-sight connections between communicating parties in the case of lasers and a continuous physical connection in the case of fiber optics.[92] This limits—although hardly eliminates—their usefulness. Such capabilities may work in some settings; for example, submarines may be able to use lasers to transmit data to nearby surface ships.[93]

GPS has been a major boon to military operations in recent years. The United States plans to increase the strength of its military GPS signals with new satellites within a few years—at which point it will no longer intentionally degrade the accuracy of civilian signals (which are to be carried on separate frequencies). Localized jamming might still degrade the accuracy of precision weapons using the military's enhanced GPS. But if the weapons

88. See Defense Science Board 1996 Summer Study Task Force, *Tactics and Technology for 21st Century Military Superiority*, vol. 1 (Department of Defense, 1996).

89. Hanna, "Task Force XXI: The Army's Digital Experiment," pp. 1–4; Scott R. Gourley, "U.S. Glimpses a 'Digitized' Future," *Jane's International Defense Review*, vol. 30 (September 1997), p. 55.

90. Nicholas Negroponte, *Being Digital* (Alfred A. Knopf, 1995), p. 16.

91. Cooper, *The Emerging Infosphere*, p. 49.

92. A current laser technology designed for infantry soldiers has a range of one to two kilometers; much greater ranges are possible, but only for individuals and weapons with a direct line of sight between them. See William D. Siuru, "Technology 'That Is Available Now,'" *Marine Corps Gazette*, vol. 81 (January 1997), pp. 29–33.

93. Sherry Sontag and Christopher Drew, with Annette Lawrence Drew, *Blind Man's Bluff* (New York: Public Affairs, 1998), p. 270.

have a backup terminal inertial guidance capability, they will be able to strike stationary targets with fairly good accuracy, even if GPS guidance is lost. Antisatellite weapons will, for the foreseeable future, have a difficult time shooting down the large numbers of satellites that make up these constellations (and redundancy will generally be built into future systems, just as it is in the GPS system).[94] GPS systems may also become even more accurate; for example, if satellite data can be supplemented by low-vulnerability ground stations within a thousand-mile-wide theater of operations, GPS precision may approach the vicinity of one meter, as demonstrated at Eglin Air Force Base in 1995.[95]

Computers

The litany of remarkable statistics about progress in computer technology is familiar to most. The maximum number of computer computations per second has been increasing by a factor of ten every five years for at least a decade.[96] Personal computers have roughly doubled in speed every two years since IBM's personal computer was introduced in 1981.[97] By the early 1990s, they were already as fast as machines considered to be supercomputers in the 1980s had been.[98] In fact, computers in the mid-1990s were 100 times more powerful than had been predicted fifteen years before. Costs have been dropping quickly; for example, figure 3-2 shows that between 1985 and 1990 the cost of a given amount of computing power declined by more than a factor of ten.[99] Equally impressive, memory storage costs have

94. Scott Pace and others, *The Global Positioning System: Assessing National Policies* (Santa Monica, Calif.: RAND, 1995), pp. 45–55, 86–91; George I. Seffers, "Army War Game Reveals Power of Commercial Data," *Defense News*, September 22, 1997, p. 44. GPS receivers can also be developed to create electronic "blind spots," or null points, in their receiving patterns that greatly reduce the range of jammers. See David Foxwell and Mark Hewish, "GPS: Is It Lulling the Military into a False Sense of Security?" *Jane's International Defense Review*, vol. 31 (September 1998), pp. 33–41; Paula Shaki, "Civil GPS Changes Hinge on Military Upgrades," *Defense News*, May 10, 1999, p. 14.

95. Defense Science Board 1998 Summer Study Task Force, *Joint Operations Superiority in the 21st Century*, vol. 2, *Supporting Reports* (Department of Defense, 1998) p. 23.

96. Kenneth Flamm, "Controlling the Uncontrollable," *Brookings Review*, vol. 14 (Winter 1996), pp. 22–25.

97. Martin Libicki, "Technology and Warfare," in Patrick M. Cronin, *2015: Power and Progress* (National Defense University, 1996), p. 120.

98. Panel on the Future Design and Implementation of U.S. National Security Export Controls, *Finding Common Ground: U.S. Export Controls in a Changed Global Environment* (Washington, D.C.: National Academy Press, 1991), p. 257.

99. Panel on the Future Design and Implementation of U.S. National Security Export Controls, *Finding Common Ground*, p. 254.

Figure 3-2. *Evolution of Computing Power*

Computing power per unit cost (bytes/second divided by fiscal year 2000 dollars)

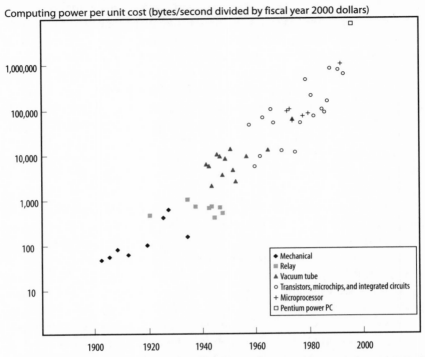

Source: Fred Levien, "Information Technology: The Plain Truth," *Journal of Electronic Defense*, vol. 22 (April 1999). Reprinted with permission of the *Journal of Electronic Defense*, Horizon House Publications. *The Journal of Electronic Defense* is the official publication of the Association of Old Crows. Full text search and retrieval of publication articles is available at http://jedonline.com.

been declining by half every two years for about twenty-five years, and memory capacity has been doubling every eighteen months.[100]

These trends show no immediate signs of slowing. It will be a while before traditional techniques for improving speed encounter constraints from the laws of physics and the speeds of light and electric currents. In practical terms, a new method to carve out circuitry on computer chips will be needed within a few years of the start of the century. That is because optical lithography, the current technique for etching, uses (visible) wavelengths of light that are too wide for the degree of miniaturization that will be needed on circuit boards to uphold Moore's "Law." That law—actually just a remarkably prescient and reliable rule of thumb—was devised by cofounder of Intel

100. On memory storage, see Kenneth Flamm, *Mismanaged Trade: Strategic Policy and the Semiconductor Industry* (Brookings, 1996), p. 10; on capacity, see Cooper, *The Emerging Infosphere*, pp. 32–35.

Gordon Moore in the 1960s. It states that the number of transistors on a semiconductor chip will double every eighteen to twenty-four months. It will be challenging to devise the technologies necessary to keep making transistors smaller; candidates include ultraviolet lithography or etching with actual electrons. But the potential for success seems good.[101] Over time, even more radical concepts may allow computing speed to keep increasing; however, even if such new techniques eventually prove elusive, military computers will be enormously better in 2020 than in 2000.[102]

Of the many benefits of improved computing speed—or, put differently, more power per cubic centimeter of machine—will be devices such as handheld computers and wrist telephones. These types of developments will continue to tie together the future battlefield and make it possible to equip every soldier with high-performance computational and communications gear—provided the means can be found to transmit data in sufficient quantities and with good reliability.[103] That capability could be of great use to soldiers in urban combat, as well as those who might be deployed in future scout teams to call in long-range firepower against enemy forces.[104]

In addition, automatic target recognition (ATR) technologies may soon make possible autonomous munitions that could be very resilient to countermeasures. ATR could help militaries cope with the inordinately large streams of data generated by imaging systems, which are reportedly at least ten times greater than what can now be processed. They could also permit much greater search rates from platforms such as UAVs that would otherwise be constrained in their searches by the need to send huge amounts of data through the airwaves for remote processing. As miniaturized sensors with good resolution—such as lidars—become available, the potential for effective ATR will grow.[105]

101. David Voss, "Moses of the Nanoworld," *Technology Review*, vol. 102 (March–April 1999), pp. 55–57.

102. New concepts could include three-dimensional chips, Josephson Junction computers, and quantum computers. See Paul C. W. Chu, "High-Temperature Superconductors," *Scientific American*, September 1995, pp. 164–65; Neil Gershenfeld and Isaac L. Chuang, "Quantum Computing with Molecules," *Scientific American*, June 1998, pp. 66–71; National Security Study Group, "Study Group Paper on Future Technology," reprinted in *Inside the Pentagon*, April 15, 1999, p. 14.

103. Richard Comerford, "Pocket Computers Ignite OS Battle," *IEEE Spectrum* (May 1998), pp. 43–48.

104. Jonathan S. Landay, "Guns that a 'Grunt' Would Love," *Christian Science Monitor*, April 6, 1998, p. 1; Jason Glashow, "U.S. Army, Marines to Study Urban Warfare Technology," *Defense News*, May 1–7, 1995, p. 18.

105. David Fields, "The Adaptive Spectral Reconnaissance Program," paper prepared for the DARPA Tech '97 Systems and Technology Symposium, Kansas City, Mo., September 1997;

The Realities of Physics—and of Countermeasures

A number of technical realities will either limit the potential of future battlefield communications or put them at considerable risk of being degraded. Moreover, even in the realm of computers, there are important caveats to what one should expect about future performance.

COMMUNICATIONS. A basic and inherent limit on communications is bandwidth. Only so much information can be carried by radio frequencies at any given time. For example, a normal radio channel spanning 6 million hertz can convey about 20 million bits per second, a data rate that allows one video image to be transmitted every tenth of a second or so. That is a rapid rate if one needs only to exchange data on the identities, positions, and speeds of enemy forces; it is a slow rate, by contrast, if one needs to send raw imagery data obtained by numerous sensors to a central base for processing.[106] These data transmission constraints highlight the importance of developing software capable of automatic target recognition.

The most basic enemy threat to communications is jamming, and it is a serious threat to GPS-aided systems as well as radio voice and data links. Various antijamming techniques are available, as noted, and overall trends suggest that a technically sophisticated country will be able to preserve robust communications. However, in network-centric warfare, the stakes are so high that this vulnerability should not be downplayed. If jamming succeeds, information networks collapse—at least locally.[107]

High-altitude nuclear bursts—which might not kill anyone directly and thus might be used by a foe with some hope of avoiding massive retaliation—would damage unprotected semiconductor chips on a wide variety of devices below. They produce a burst of radio-frequency energy lasting a fraction of a second and known as high-altitude electromagnetic pulse (HEMP). The semiconductors of twenty years ago were about 10 million

---

David A. Fulghum, "Small Recon Satellites Win 1999 Budget Funding," *Aviation Week and Space Technology*, February 9, 1998, p. 28; James Bamford, "Our Best Spies Are in Space," *New York Times*, August 20, 1998, p. A23; Jonathan Schonfeld, "Automatic Target Recognition (ATR) for Rapidly Deployable, Outnumbered Forces in Wide-Area Engagements," in Defense Science Board 1996, *Tactics and Technology for 21st Century Military Superiority*, vol. 3, pp. II-90 through II-92; Craig A. Covault, "Locass Attack System Development Advances," *Aviation Week and Space Technology*, October 26, 1998, pp. 52–55.

106. Negroponte, *Being Digital*, p. 52; Coleman Bazelon, Perry Beider, and David Moore, *Where Do We Go from Here? The FCC Auctions and the Future of Radio Spectrum Management* (Congressional Budget Office, 1997), p. 2; Jackson, *Classical Electrodynamics*, p. 291.

107. Foxwell and Hewish, "GPS."

times more sensitive to this effect than vacuum tubes; today's semiconductors are more sensitive still.[108]

At low altitudes, electromagnetic pulse (EMP) is also produced. But it falls off in strength very quickly and does not cause damage beyond a few kilometers' distance (a region in which other nuclear effects would obviously also be important). HEMP, by contrast, varies little with distance and produces a reasonably strong electric surge at any point within sight of it on earth—that is, to the horizon. The horizon is a function of the altitude of the burst. For example, a detonation at fifty miles' altitude would affect a circular area directly below with a radius of about 600 miles; a single burst in the hundreds of kilotons range at 200 miles' altitude could affect almost the entire United States. HEMP is caused by the effects of gamma rays, which travel at the speed of light and ionize molecules they encounter. Since they encounter more molecules going down than going up (due to the fact that the atmosphere is denser at lower altitudes), and since the electrons produced by the ionizations move much more quickly than the positive ions, the net effect is to create an electric field somewhat akin to what would result from lightning.[109]

Most vulnerable to HEMP are many commercial electronics devices, which are popular in defense circles because they are inexpensive but which are less resilient than most tailored defense electronics. The Pentagon has reduced its emphasis on radiation hardening considerably. Because shielding must generally be tailored to the specific size and shape of a given object, it is difficult and expensive to harden objects against HEMP.[110] As such it works at cross-purposes with the popular notion that modern electronics—as well as commercial off-the-shelf technologies—will enable inexpensive improvements in defense capabilities. Pentagon expenditures for hardened microchips have dropped from about $1 billion to $100 million annually, creating the potential for major vulnerabilities in the future U.S. armed forces.[111]

Some vehicles hardened to withstand high-altitude EMP—like fighter jets and tanks—might emerge unscathed, but those carried by hand or in jeeps

108. As George Ulrich of the Defense Special Weapons Agency put it, "Each new generation, smaller and needing less power, exacerbates those vulnerabilities." See Joseph C. Anselmo, "U.S. Seen More Vulnerable to Electromagnetic Attack," *Aviation Week and Space Technology*, July 28, 1997, p. 67.

109. Glasstone, *The Effects of Nuclear Weapons*, pp. 514–29.

110. Glasstone, *The Effects of Nuclear Weapons*, pp. 527–29; Sean J. A. Edwards, "The Threat of High Altitude Electromagnetic Pulse to Force XXI," *National Security Studies Quarterly*, vol. 3 (Autumn 1997), pp. 61–80.

111. Mark Walsh, "DoD Tries to Boost Nuke-Resistant Chip Industry," *Defense News*, May 26–June 1, 1997, p. 3.

and trucks and not specially hardened could be destroyed. The only modern satellite communications system built to be well hardened against nuclear attack, and to overcome the effects of nuclear bursts on atmospheric propagation of radio waves, is the MILSTAR system. Other satellite systems tend to be rather vulnerable.[112]

Radio-frequency (RF) weapons could cause similar effects over a smaller distance. They produce very short pulses, lasting only billionths of a second, with very high power—often billions of watts.[113] Whereas an adversary would damage its own unshielded electronics almost as much as an enemy's with high-altitude EMP, radio-frequency weapons have more localized effects. Like the situation with HEMP, chips are less able to withstand RF weapons than are vacuum tubes. These weapons are in their infancy but are likely to become available—and to more countries than the United States—in the next two decades. Russia is known to have capability in this area already, and the basic physics is not complicated. There are reports that such weapons have been tested in the United States as well. Their range appears to be tens of meters in a man-portable form, perhaps hundreds of meters if vehicle-sized; microwave lasers, or masers, could have ranges ten times greater than that.[114] According to the 1999 study by the special U.S. House of Representatives panel led by Congressman Christopher Cox, China may have acquired some information about U.S. research on such electromagnetic weapons.[115]

As a counter to these dangers, radios and other electronics could be better hardened. Alternatively, they could be used less frequently so that they would tend to be off—with antennas somehow delinked from the electronic guts of the devices—most of the time, limiting their vulnerability to attack. Research is being pursued to reduce the vulnerabilities of devices in these and other ways. But the short-term trend may be to increase, rather than mitigate, such vulnerabilities. Not only are hardened chips becoming less common, but some of the composite materials used in modern aircraft and

112. Special Report, *Jane's Defence Weekly*, August 26, 1998, p. 25.

113. Ira W. Merritt, U.S. Army Space and Missile Defense Command, "Radio Frequency Weapons and Proliferation: Potential Impact on the Economy," statement before the Joint Economic Committee, 105 Cong. 2 sess. (February 25, 1998).

114. Barbara Starr, "Russian Bomb-Disarming Device Triggers Concerns," *Jane's Defence Weekly*, March 18, 1998, p. 4; Carlo Kopp, "The E-Bomb—a Weapon of Electrical Mass Destruction," Monash University (Australia), 1998; "Russian Electronic Bomb Tested in Sweden," *Agence France-Presse*, January 21, 1998; David A. Fulghum, "Microwave Weapons Await a Future War," *Aviation Week and Space Technology*, June 7, 1999, pp. 30–31.

115. Robert Wall, "Panel Says Spying Aided Conventional Weaponry," *Aviation Week and Space Technology*, May 31, 1999, p. 31.

armored systems would protect electronics within the weapons systems less well than traditional metal skins.[116]

Another reality of communications technology that might not help the United States is the proliferation of satellites. Because of it, other countries will develop or acquire reliable and redundant communications capabilities, and the United States will have a difficult—if not impossible—time physically eliminating all such adversarial capabilities (unless it is willing to use exoatmospheric nuclear bursts to destroy swaths of satellites, including many of its own).[117] By 2012, for example, the U.S. military reportedly expects that 600 to 700 civilian and military satellites will be in orbit. War games set in that time period, to the extent they are able to be technologically prescient and accurate, corroborate the judgment that adversarial satellite communications will endure even after use of antisatellite weapons by the United States.[118]

COMPUTERS. HEMP and radio-frequency weapons can affect computers, too. They damage all electronics, even if they typically make their way into computers through communications devices or other conductors.

As noted, RMA proponents often emphasize the tremendous capabilities that future computer technology will provide in the form of automatic target recognition. Having machines process imagery data to determine whether targets are present will surely be of great benefit. Indeed, at a certain point ATR is necessary, since the limited amounts of radio bandwidth in the electromagnetic spectrum may preclude sending raw data from many imaging platforms to central processing centers and necessitate that data be processed by the platform acquiring it.[119] However, false alarm rates for ATR systems now in development are fairly high. Countermeasures, like those mentioned previously, may keep false alarm rates high even as ATR capabilities improve. Unless military vehicles travel in a convoy or tactical formation, or are given away by their radio communications, they may not be recognized even if they are seen. ATR may help one know that a mili-

116. James F. O'Bryon, Office of the Secretary of Defense, statement before the Joint Economic Committee, 105 Cong. 2 sess. (February 25, 1998); Merritt, "Radio Frequency Weapons and Proliferation."

117. Lisa Burgess, "Sunspots Could Wreak Havoc with Satellites Next Year," *Defense News*, July 6–12, 1998, p. 11.

118. William B. Scott, "Wargame Shows Impact of Air/Space Action," *Aviation Week and Space Technology*, December 8, 1997, pp. 26–27.

119. As noted, imaging systems reportedly create streams of data at least ten times greater than what can now be processed; that amount is sure to grow greatly as UAVs and other sensor platforms spread. See Bamford, "Our Best Spies Are in Space."

tary vehicle *might* be present, but it may not be able to distinguish it from a civilian vehicle.[120]

Hacking and computer viruses will always cause some concern that key networks will be unavailable at a key moment. A wide range of not only military but also key civilian systems are currently vulnerable to cyber sabotage. In fact, a number have already been attacked, including eleven defense networks in early 1998 and a naval intelligence unit in 1999.[121] These threats are not serious in all cases. A large number of the Pentagon's 100,000 or so networks would have only a small or negligible role in warfighting operations. Were the integrity of some—or even many—of them compromised, the immediate consequences would be minor for security purposes (even if there was inconvenience associated with restoring their integrity).[122]

Moreover, the Pentagon can take defensive steps to protect its most critical computer systems. Better fire walls can be created. Even though few—if any—are impervious, more users can be kept out and greater effort will be required to penetrate any given system.[123] In addition, key information can be routed through a limited-access or secure internet for the military (and similar closed-loop command systems for key civilian infrastructure), meaning that hacking or virus implantation could be carried out only by someone within the network.[124] As another line of defense, redundant programs and networks can be developed for particularly critical computing capabilities, such as those collecting and analyzing targeting data, transmitting communications, or keeping track of supplies.[125] The United States

120. For example, one effort to date using state-of-the-art infrared technology produced more than twice as many false alarms as actual targets, while also missing about 20 percent of the targets. For descriptions of this and other current research in this area, see Brian Ernisse and others, "Complete Automatic Target Cuer/Recognition System for Tactical Forward-Looking Infrared Images," *Optical Engineering*, vol. 36 (September 1997), pp. 2593–603. Also, see Matthew R. Whiteley and others, "Detection of Military Vehicles Using Infrared Spectral Radiometric Signatures," *Optical Engineering*, vol. 35 (December 1996), pp. 3531–548; Defense Science Board 1998, *Joint Operations Superiority in the 21st Century*, vol. 2, p. 28.

121. President's Commission on Critical Infrastructure Protection, *Critical Foundations: Protecting America's Infrastructures* (1997); Bradley Graham, "11 U.S. Military Computer Systems Breached by Hackers This Month," *Washington Post*, February 26, 1998, p. A1; Bill Gertz, "Pentagon Fortifying Computer Networks to Stymie Hackers," *Washington Times*, April 17, 1998, p. A3; Richard Lardner and Pamela Hess, "Pentagon Looks for Answers to Massive Computer Attack," *Defense Information and Electronics Report*, February 13, 1998, p. 1; "One on One with Arthur Money," *Defense News*, January 18, 1999, p. 22; Matthew Campbell, "Russian Hackers Steal US Weapons Secrets," *Sunday Times* (London), July 25, 1999.

122. Gertz, "Pentagon Fortifying Computer Networks to Stymie Hackers."

123. Bruce D. Nordwall, "Cyber Threats Place Infrastructure at Risk," *Aviation Week and Space Technology*, June 30, 1997, p. 51.

124. Nordwall, "Cyber Threats Place Infrastructure at Risk."

125. Glenn C. Buchan, "Implications of Information Vulnerabilities for Military Opera-

also could hardwire key data into more of its computing so that computers could simply be rebooted, expunging foreign information. Rather than requiring programming against every possible virus, computers may become capable of recognizing foreign data or programs and attacking them automatically—much as the immune system protects the human body.[126]

Still, risks would remain. For example, closed-loop networks could be sabotaged by insiders. If they were sabotaged in a way that would cause systems to crash only in wartime, they could be especially dangerous.[127]

Even if computers are hardened against EMP and secured against viruses, their innate potential is limited. Computers can do only so much; they are manipulators of data, but they do not create the data or make direct use of it to themselves destroy enemy systems. They are positioned, in large part, in an intermediate location in the military technology food chain.

A telling example of what computers can—and cannot—do involves the design and maintenance of nuclear weapons. The Department of Energy's current stockpile stewardship program is intended to use advanced experimental techniques to improve understanding of the physics of nuclear warheads without the need for actual nuclear detonations. It is also trying to develop extremely sophisticated computer models that can be calibrated with the real-life data obtained from past nuclear explosions and then used to predict warhead performance. Recent breakthroughs in computing power have made possible machines that can capture the complexity of a three-dimensional process, rather than use an imperfect two-dimensional proxy for a real warhead as has traditionally been the case. Those models are intended, among other things, to help weapons scientists monitor the effects of aging on warheads to know when to remanufacture key parts.

However, those computer programs and computations will be only as good as the physical models and test data used within them. Most weapons scientists consider even advanced computing capabilities no more than a partial substitute for nuclear testing. Few if any believe they will be able to reliably design sophisticated new warheads with computer models alone. Nor is the concern confined to the issue of building fundamentally new warheads. Problems could arise if new materials are used in existing warhead

---

tions," in Zalmay M. Khalilzad, A. H. Marshall, and John P. White, eds., *The Changing Role of Information in Warfare* (Santa Monica, Calif.: RAND, 1999), pp. 290–95.

126. Gloria Wilt, "Making Information Safe," *Science and Technology Review*, Lawrence Livermore National Laboratory (January–February 1998), pp. 4–11; George I. Seffers, "U.S. Readies Vaccine to Fight Virtual Virus," *Defense News*, September 22–28, 1997, p. 1.

127. David A. Fulghum, "Cyberwar Plans Trigger Intelligence Controversy," *Aviation Week and Space Technology*, January 19, 1998, p. 55.

designs in the future. For example, if in the future plutonium is cast, rather than machined, to its proper size and shape—or if a different type of high explosive is used to initiate the compression of the plutonium in a nuclear weapon's so-called primary—even an extremely fast computer may prove incapable of properly modeling the interactions that would result. If the physical processes that would occur in a real explosion with those new materials are not well understood, the computer will not be able to compensate. This is not an argument against the nuclear test ban treaty, since the United States can always retain some types of warheads periodically rebuilt to original engineering specifications, or build a new type of simple, robust warhead without the benefit of testing as an insurance policy of sorts (albeit at some degradation to the power of the warhead). However, it does serve to remind one of the limitations of computers.[128]

This is true not only for U.S. nuclear weapons, of course, but also for those of other nuclear powers, fledgling nuclear-weapons states, and potential proliferators. As one Department of Energy physicist described the effects of better and speedier computers: "If your [input] data are wrong, you just get the wrong answers faster."[129]

Conclusion

Progress will continue to be very rapid in computer and communications technologies. Computers will continue to become faster, cheaper, lighter, and more widely used on the battlefield. Factors of ten or more in improvements in speed and memory capacity are likely between now and 2020, though some major technological hurdles will have to be surmounted if performance is to improve by much more than tenfold by then. Rapid dissemination of targeting data between various platforms will, at least in the absence of enemy countermeasures, allow real-time network warfare. Weapons platforms will no longer need to acquire a target themselves before shooting and will not need to wait hours—or even minutes—for the coordinates of many types of enemy systems, particularly large, mechanized systems.

However, important limitations and vulnerabilities will remain even in these technological spheres. For example, certain high-density data trans-

128. David Mosher, *Preserving the Nuclear Weapons Stockpile under a Comprehensive Test Ban* (Congressional Budget Office, May 1997), pp. 19–20; see also Michael O'Hanlon, *The Bomb's Custodians* (Congressional Budget Office, July 1994).

129. Carla Anne Robbins, "Why Nuclear Threat Today Can Be Found at the Electronics Store," *Wall Street Journal*, December 14, 1998, p. A1.

Figure 3-3. *Projected Advances in Key Military Technologies, 2000–20*[a]

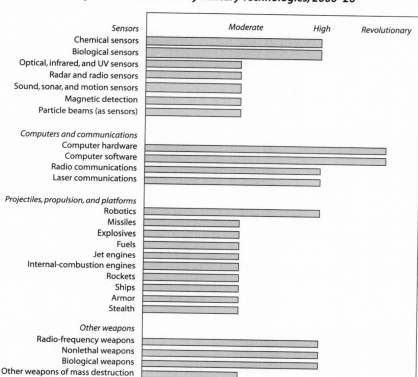

Source: Author's estimates.

a. The terms moderate, high, and revolutionary are subjective and somewhat imprecise. In general terms, technologies showing moderate advances might improve their performance by a few percent or at most a couple of tens of percent—in terms of speed, range, lethality, or other defining characteristics—between 2000 and 2020. Those experiencing high advances will be able to accomplish tasks on the battlefield far better than before—perhaps by 50 to 100 percent, to the extent improved performance can be so quantified. Finally, technology areas in which revolutionary advances occur will be able to accomplish important battlefield tasks that they cannot now even attempt.

missions, such as the widespread use of video imagery over the battlefield, will probably not be possible due to the inherent limitations of radio waves to carry information. Even more limited transmissions of information, the ability of computer-dependent aircraft to stay airborne, and the ability of ground stations to survive attack may be seriously degraded by radio-frequency weapons and high-altitude nuclear-induced electromagnetic pulses.

More fundamentally, whatever the pace at which computers and communications systems are advancing—and the pace is admittedly impressive—we must not overestimate what these capabilities can themselves accomplish. They are, after all, military intermediaries that take information from sensors and then process and disseminate it to weapons. Good computing and communications can fully exploit the potential of existing sensors and weapons, but they cannot themselves find or destroy the enemy.

As for sensors, the overall technical prognosis is good but far less impressive on the whole, as can be seen from figure 3-3. Developments in military sensors will provide U.S. armed forces with a number of new options and capabilities between now and 2020. For example, the miniaturization of chemical weapons detectors, taken together with improvements in robotics, will allow earlier warning of the presence of chemical munitions on the future battlefield. These capabilities will also permit U.S. troops to search limited areas for signs of chemical weapons without putting themselves directly in harm's way. Similar types of synergies between robotics and detectors will allow unmanned underwater vehicles to search for mines or enemy submarines in dangerous areas like the littorals of southwest Asia. Miniature aerial vehicles and unmanned ground stations will improve the ability of U.S. forces to track exposed enemy forces in urban and infantry environments. Snipers will often be detected and counterattacked after their first shot, thanks to new acoustic and laser detectors. Foliage-penetrating radars may help U.S. forces look beneath jungle canopy to detect concentrations of enemy units. Larger numbers of imaging satellites as well as platforms like the JSTARS radar-imaging aircraft may make it easier to not only detect, but also continuously track, large military objects on the battlefield. In time, autonomous radars on air-to-ground missiles may become practical, making precision attack possible even under poor weather conditions.

At least as noteworthy as potential sensing capabilities, however, is what will *not* become possible. For example, in infantry combat settings, no sensors will be able to distinguish unarmed combatants from civilians (except in rare cases in which the voice or DNA signature of a specific individual is known). Even armed combatants will be hard to identify from any distance in most cases. It will remain possible for low-tech adversaries of the United States to hide mortars or small missiles in trucks, keep secret arms caches deep in the interiors of large buildings or underground, and ambush even technologically superior enemy military units.

Underwater targets will remain difficult and time consuming to find. Military vehicles built to resemble civilian vehicles in their external shapes, structural materials, and engines will be difficult to identify in many settings. They may or may not be detectable in forests or other complex terrain. Some of the sensors that might be used to find them in these types of settings—such as foliage-penetrating radar—must overcome enormous technical hurdles to be feasible. Other sensors, such as acoustic and magnetic devices, have very short ranges and would need to be deployed in the right general vicinity of an enemy movement before it occurred.

It will almost surely remain impossible for advanced sensors to locate fissile materials or nuclear weapons, particularly those with even a modicum of shielding, from a distance greater than several hundred meters or—at most—a few kilometers. Similar conclusions apply to chemical and biological weapons stored inside sealed containers, artillery shells, or missile warheads.

In summary, RMA proponents are correct to emphasize that an information grid with real-time data processing and dissemination can synergistically integrate sensors, vehicles, and weapons to produce impressive new military performance. In fact, as argued further in chapter 7, the time does seem right for the United States to emphasize such a system of systems philosophy in its weapons acquisition efforts, purchasing advanced computers and communications systems (as well as electronics-dependent technologies such as advanced munitions) in ample numbers. Electronics, computers, and communications systems are progressing rapidly—in fact, it is fair to use the adjective revolutionary to describe advances within these specific realms of defense technology.

However, the popular notion of information dominance, voiced by most RMA proponents and repeated in the Pentagon's *Joint Vision 2010*, simply goes too far, as does the concept of dominant battlespace knowledge. It is one thing to determine that a silicon-based modernization approach is the most cost-effective way to improve U.S. defense capabilities at this point in history. It is quite another to claim that reconnaissance strike complexes will radically transform warfare in the years ahead, necessitating wholesale changes in U.S. security policy and defense resource allocation in the process. The former conclusion seems a sober analytic judgment. The latter is a leap of faith that seems to blithely assume breakthroughs in sensor technology upon which a sober scientific survey casts considerable doubt.

CHAPTER FOUR

Vehicles, Ships, Aircraft, and Weaponry

What about technical trends in systems such as tanks, ships, jets, and missiles? Chapter 3 focused primarily on electronics and sensors, which are clearly important in major combat weapons and platforms. But the performance of these weapons and platforms also depends heavily on their propulsion systems, armor, and related characteristics. These mechanical, chemical, and structural technologies are surveyed in this chapter. Nonlethal and biological weapons are also considered, as are defense systems against ballistic and cruise missile attack.

The approach taken in this chapter differs from that of chapter 3. Here, it is generally less useful to rely on the fundamental constraints and realities of physics. Instead, this chapter attempts to discern trends in various types of technology and extrapolate their future capabilities and uses. Its approach is therefore more impressionistic and more subject to error than that of chapter 3. But it is far more systematic than relying on anecdotes, pure conjecture, and false analogies with trends in computers and electronics—as many proponents of a contemporary revolution in military affairs (RMA) tend to do.

Although this chapter does not downplay the pace of progress in military technology, it does highlight a number of sobering themes for consideration by those purporting to discern an RMA on the horizon. Most important, the chapter argues that most developments in propulsion and the basic designs and dynamics of vehicles, ships, and airplanes are clearly occurring at incremental and evolutionary—not revolutionary—rates. This prognostication applies to rates of advance in the speed of ships and airplanes, the efficiency of rockets, the basic functioning of the internal-combustion engine, and the power of explosives.

These findings do not by themselves disprove the possibility of a near-term RMA. A number of modest new capabilities combined with a couple of key breakthroughs in defense technologies, all integrated in a real-time information network, may indeed make possible an RMA, at least in theory. That possibility is considered in chapter 5. In other words, the specific case-by-case technology assessments of this chapter do not rule out the possibility that the broad conclusions offered by RMA proponents, as laid out in chapter 1, may be right.

This chapter does provide grounds for skepticism, however. Taken in conjunction with chapter 3, it strongly suggests that it will be much harder to increase greatly the speed, lethality, and rapid deployability of U.S. combat forces than documents ranging from *Joint Vision 2010* to the National Defense Panel's 1997 report to a number of other RMA proposals often claim. In light of this, the second and third technical premises of the RMA movement, as spelled out in chapter 1, appear to be exaggerations at best and downright inaccuracies at worst.

Aircraft

Jet engines have improved considerably over the past several decades, becoming more efficient, powerful, and reliable. For example, in the period of roughly two decades, between the development of the F-15 and the F-22 aircraft, engine technology improved so much that the latter fighter has roughly twice the power of the former. Among the reasons for this improved performance are engine materials that allow much higher temperatures within the turbines than previously possible—around 3,400 degrees Fahrenheit, in contrast to temperatures 1,000 degrees lower in previous-generation aircraft. At the same time, the F-22 and F-15 are similar in weight, both in the aircraft as a whole and in the engines specifically—a combination that makes for a significant improvement in performance capability for the newer aircraft.[1]

1. Bill Sweetman, "The Progress of the F-22 Fighter Program," *Jane's International Defense Review*, quarterly report no. 1 (1997), p. 8. The improvements in engine technology have largely to do with the bypass ratio of air flowing through the jet engines; much more air flows through modern designs. Other efforts focus on using engine materials capable of tolerating higher temperatures, such as silicon carbide and titanium. See Eugene E. Covert, "Evolution of the Commercial Airliner," *Scientific American,* September 1995, p. 112; and Lane Pierrot, *A Look at Tomorrow's Tactical Air Forces* (Congressional Budget Office, 1997), p. 38.

Figure 4-1. *Aircraft Speed since World War II[a]*

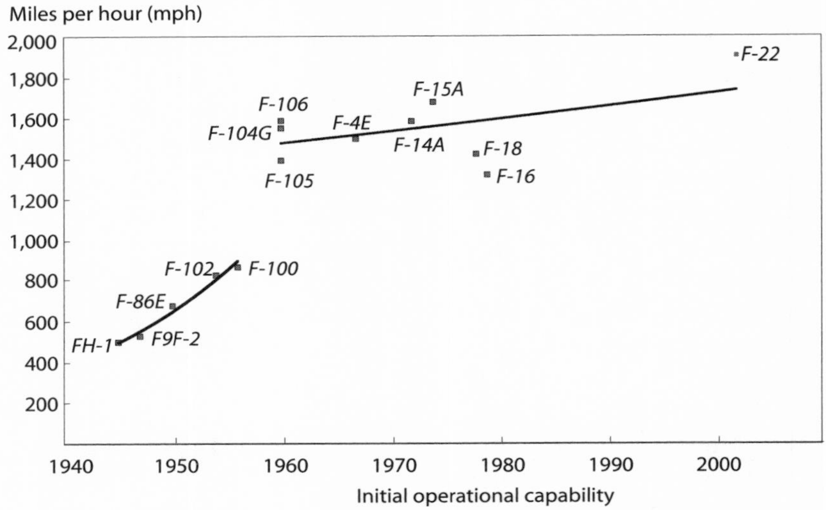

Miles per hour (mph)

Source: Enzo Angelucci, *Rand McNally Encyclopedia of Military Aircraft: 1914 to the Present* (Harrisburg: Crescent Books, 1990), pp. 398–423.

a. Mach 1 (speed of sound) = 740 mph at sea level. Aircraft speeds indicated are measured at optimum flying altitude and weight, which differ for each aircraft. The maximum estimated speed for the F-22 remains classified. The prototype YF-22 has achieved Mach 1.58 (1,203 mph) at 30,000 feet without afterburning. Using available open-source data, the author estimates the maximum speed of the F-22 at Mach 2.5 (1,903 mph) with afterburning.

(Figure 4-1 presents a longer-term perspective on improvements in combat aircraft speed.) Yet this improvement in performance is far below the rate of innovation in computers, where doublings of capability typically occur every eighteen to twenty-four months.

New aircraft materials include titanium and a number of composites, some of which have the added benefit of stealthiness. These materials will constitute about two-thirds of the total mass of the F-22, reducing the aircraft's weight by about 25 percent, as compared to an all-aluminum frame. The benefits of such materials are considerable, yet a 25 percent reduction is better thought of as evolutionary than revolutionary change.[2]

It is illuminating to think of how these types of changes in aircraft compare with those of the interwar years, when almost all would agree that a revolution in military affairs truly did occur. As Andrew Krepinevich has emphasized, the U.S. Navy experimented with many different types of carriers, not knowing which would prove most effective. Wisely, it developed

2. Sweetman, "The Progress of the F-22 Fighter Program," p. 6.

three classes of carriers but produced a total of only four ships. By contrast, the British Royal Navy put most of its eggs in one basket—that of small carriers. As aircraft improved, larger carriers became necessary to accommodate them.[3] In the last several decades, however, it has not been necessary to change the size of carriers, and future carrier aircraft will not demand it either. Today, basic propulsion systems and designs for aircraft, ships, and internal-combustion vehicles are changing much more gradually than in the early twentieth century, when two of those three technologies had only recently been invented. As Martin Libicki put it in regard to the specific example of jet engines, changes are now of degree rather than of kind.[4]

Breakthoughs in stealth technology over the last three decades have been quite impressive, as suggested in figure 4-2 and table 4-1. Progress continues to make aircraft less observable, not only against radar but also against other types of sensors. Modern aircraft like the F-22 have conformal nozzles, high-bypass engines that keep exhaust temperatures low by mixing hotter air with cooler air before expunging the mix, and various types of paints and surface coatings that absorb solar infrared radiation and limit infrared emissions from friction-generated heat. The goal with these types of technologies is to ensure that infrared signatures constitute no more of a vulnerability for stealth aircraft than do their radar signatures.[5]

Helicopters are becoming more stealthy, too. For example, the Comanche helicopter is to have only one-fourth the infrared signature, one-half the acoustic signature, and less than one-hundredth of the radar signature of current helicopters like the Apache. The radar cross section of the Comanche is thus probably less than one square meter—quite modest, albeit greater than the 0.01 square meter range that probably typifies the F-117, B-2, and F-22 fixed-wing aircraft.[6] (The joint strike fighter, or JSF, reportedly has just as small, or even smaller, a head-on radar cross section, though it may not wind up being quite as stealthy from certain side and rear angles as the B-2 and F-22. In addition, the JSF is reportedly being designed to have small radar reflectivity even against lower-frequency radars.[7])

3. Andrew F. Krepinevich Jr., testimony before the House National Security Committee, Subcommittees on Military Procurement and Research and Development, 105 Cong. 2 sess., October 8, 1998 (Washington, D.C.: Center for Strategic and Budgetary Assessments, 1998), pp. 6–7.

4. Martin Libicki, "Technology and Warfare," in Patrick M. Cronin, *2015: Power and Progress* (National Defense University, 1996), p. 120.

5. Sweetman, "The Progress of the F-22 Fighter Program," pp. 9, 17.

6. Stanley W. Kandebo, "U.S. Army Ponders Comanche Restructure," *Aviation Week and Space Technology*, June 1, 1998, pp. 26–27.

7. David A. Fulghum, "JSF Reflection Is Golf-Ball Sized," *Aviation Week and Space Technology*, February 15, 1999, p. 27.

Figure 4-2. *The Benefits of Stealth*

Radar cross section (notional units)[a]

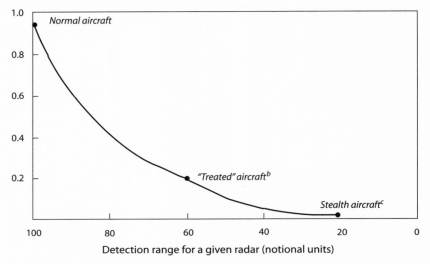

Detection range for a given radar (notional units)

Sources: Adapted from Office of Naval Intelligence, *Worldwide Challenges to Naval Strike Warfare* (Office of Naval Intelligence, January 1996), p. 18; David A. Fulghum, "Secret Upgrades Target Stealthy Cruise Missiles," *Aviation Week and Space Technology*, August 24, 1998, pp. 22–23.

a. Roughly speaking, the scale for radar cross sections corresponds to square meters for fighter aircraft. From certain angles, however, nonstealthy aircraft such as the F-15 can have radar cross sections of between 5 and 10 square meters.

b. The exterior surfaces of "treated" aircraft have been slightly modified to reduce radar cross sections. An example might be the F/A-18E/F Super Hornet.

c. An example might be the F-22 Raptor.

Progress with transport aircraft is occurring at a modest pace. The V-22 Osprey tilt-rotor aircraft is an impressive new capability. With a top speed of more than 300 miles per hour (roughly 500 kilometers per hour), it is at least 50 percent faster than modern helicopters. With a normal mission radius of more than 200 miles (more than 300 kilometers), it also has greater range than most current helicopters (though much less of an advantage than the Marine Corps routinely claims when comparing it against one particular shorter-range system).[8] The Osprey is not particularly fast, large, efficient, or stealthy, however. Its reduced vulnerability to enemy action has often been overstated. It is notably better than modern helicopters equipped with

8. Marine Corps briefing on V-22 Osprey, June 13, 1997.

Table 4-1. *Visibility of Selected Objects to Radar*

Object	Radar cross section (square meters)	Typical SAM tracking range[a] (kilometers)
Large transport aircraft	100	250
Conventional fighter plane	10	210
B-1B bomber	1	125
Cruise missile	0.1	80
Large bird (or B-2, F-22?)	0.01	50
Large insect (or B-2, F-22?)	0.001	35

Source: Congressional Budget Office, *A Look at Tomorrow's Tactical Air Forces* (January 1997), p. 77.
a. SAM = surface-to-air missile.

similar countermeasures only when being shot at by small arms.[9] Thus it is an important new technology but probably not a revolutionary one.[10]

The Pentagon hopes to field a transport helicopter by 2015 or so that would require 40 percent less fuel, weigh 60 percent less, and travel about 50 percent faster than existing models. It would certainly be nice to achieve those goals. However, they are precisely that—goals, not realities—and actual aircraft could well fall short of them. Even if attained they are not particularly striking for aircraft that will fly in the 2020s and beyond. The Pentagon clearly expects helicopter transport in the early decades of the twenty-first century to be better than, but in the end similar to, helicopter transport of the late twentieth century.[11]

Fixed-wing transport aircraft may be able to fly more efficiently by 2020 by reducing the effects of turbulence on their airframes. A plethora of microsensors and small flaps or holes in the wings connected to suction pumps may be used to detect turbulence and selectively modify airflows to reduce it.[12] Large tilt-rotor aircraft may someday be possible for long-range strategic transport missions, but these technologies lie well in the future.[13] In addition, even if perfected, they will not change the fact that airplanes will possess only modest payloads by comparison with the weights of major

9. Assistant Secretary of Defense David S. C. Chu, comments at a special hearing before a subcommittee of the Senate Committee on Appropriations, 101 Cong. 2 sess., July 19, 1990 (Government Printing Office, 1990), p. 51; L. Dean Simmons, "Assessment of Alternatives for the V-22 Assault Aircraft Program," Institute for Defense Analyses, June 1990, Alexandria, Va., p. 17.

10. Chu, comments before a subcommittee of the Senate Committee on Appropriations, p. 47.

11. George I. Seffers, "Rotorcraft Design to Target All U.S. Services," *Defense News*, May 17, 1999, p. 14.

12. See Covert, "Evolution of the Commercial Airliner," p. 113.

13. Colin Clark and David Mulholland, "Pentagon Science Chief Pushes Leap-Ahead Research," *Defense News*, November 9, 1998, p. 6.

ground force units (see the subsequent section on ground vehicles for prognostications about their future size and weight).

It is possible that, by 2020 or sooner, blimplike aircraft will once again
be used for transport. Even if quite slow by airplane standards, they could
be very fast by ship standards, crossing the Atlantic in a day or so at speeds
of more than 100 miles per hour. One design, which is actually half blimp
and half airplane, would carry about 1 million pounds, or 500 tons, of
payload—roughly ten times that of current transport aircraft. It would allow
equipment to be rolled on and off as in a dedicated military sealift vessel.
This concept is intriguing and could change the nature of long-range military transport significantly; however, the practicality, cost, and vulnerability of such systems cannot yet be determined.[14]

Many of the most exciting developments in aviation are in the realm of
unmanned aerial vehicles (UAVs). They are becoming quite widespread in
the U.S. military, as indicated in table 4-2. UAVs are notable not for breaking aerodynamic performance barriers, but for allowing various missions
to be conducted at less risk to pilots.[15] Perhaps ten were shot down during
NATO's air war against Serbia in 1999, for example, and twenty were lost
in all during an air campaign in which only two NATO-manned jets and
two helicopters were lost. Their roles in Operation Allied Force reportedly
included conducting various forms of reconnaissance; they may also provide low-altitude designation for laser-guided bombs so that the bombs could
be used in some situations despite cloud cover.[16] Experimental systems are
being built for the navy that would be capable of vertical takeoff and landing, whether through tilt-rotor technology, a traditional helicopter rotor, or
a more exotic design.[17] UAVs eventually may be used to jam or attack enemy
air defenses and ultimately even to conduct traditional bombing operations.[18] Doing so might save not only pilot lives but money; the Defense
Advanced Research Projects Agency (DARPA) estimates that acquisition and
operating costs could be cut by more than 50 percent, relative to manned
aircraft carrying out the same mission.[19]

14. David A. Fulghum and Robert Wall, "Mammoth Half-Blimp Seen as Future Airlifter,"
Aviation Week and Space Technology, February 22, 1999, p. 26.

15. Michael A. Dornheim, "Global Hawk Begins Flight Test Program," *Aviation Week
and Space Technology*, March 9, 1998, pp. 22–23.

16. Nick Cook, "War of Extremes," *Jane's Defence Weekly*, July 7, 1999, p. 22.

17. Michael A. Dornheim, "Navy Moving to VTOL to Replace Pioneer," *Aviation Week
and Space Technology*, June 8, 1998, pp. 49–51.

18. Defense Science Board 1998 Summer Study Task Force, *Joint Operations Superiority
in the 21st Century* (Department of Defense, 1998), pp. 32–35.

19. Frank Fernandez, Director, Defense Advanced Research Projects Agency, statement

Table 4-2. *Current and Planned U.S. Unmanned Aerial Vehicles[a]*

Characteristics	Pioneer	Hunter	Outrider	Tier II, Predator	Tier II+, Global Hawk	Tier III-, Dark Star
Operating altitude (km)	4.6	4.6	1.5	4.6	15.2–19.8	15.2
Maximum endurance (hours)	5	11.6	4 (+ reserve)	35	38 (20 at 5,556 km)	12 (8 at 926 km)
Radius of action (km)	185	267	200	740	5,556	926
Maximum speed (km/hr)	204	196	222	204–215	639	556
Cruise speed (km/hr)	120	165	167	120–130	639	556
Sensors[b]	EO or IR	EO and IR	EO and IR[c]	EO, IR, and SAR	EO, IR, and SAR	EO or SAR
System composition[d]	5	8	4	4	...	...
Number of systems	9	7	6	12	5[e]	4[e]

Source: Defense Airborne Reconaissance Office, *UAV Annual Report FY 1997* (Department of Defense, November 1997).

* Not yet determined.

a. All systems must also include various ground stations that are not mentioned here.

b. EO = electro-optical; IR = infrared; SAR = synthetic-aperture radar.

c. SAR growth potential.

d. Number of aircraft per "system."

e. Planned.

Over the longer term, microrobotic technologies may permit very small UAVs to be employed in combat. UAVs with dimensions on the order of six inches have been developed; their possible uses include photography, target designation, jamming, dispersing motion sensors, and carrying chemical and biological sensors. It will be difficult to use traditional fixed wings at dimensions much smaller than roughly three inches, so movable or "flapping" wings may be needed if such tiny sizes are to be attained.[20] Challenges abound, however. Accomplishing the requisite aerodynamic engineering may prove infeasible. Developing small power sources is another major hurdle, though microelectromechanical systems (MEMS) may eventually provide a solution with microturbines or microjets. Even if developed, small UAVs may be so slow and vulnerable to wind and precipitation that their outdoor uses will be confined to calm days.[21]

Other countries will develop increasingly capable UAVs, as well as improved cruise missiles, in the years ahead as well. Building stealthy aircraft is difficult. Large testing facilities known as anechoic chambers, not to mention mastery of composite materials, are required; so far, even advanced U.S. allies have been unable to approach American capabilities. However, the simplicity and smaller scale of devices like cruise missiles make their radar returns easier to model with computers as well as cheaper to build.[22]

Rockets and Hypersonic Vehicles

In rocket technology, progress has probably been slower than in jets. For example, the Trident I and II missiles have the same ratio of payload to initial rocket mass, despite the fact that the second was developed more than

before the Subcommittee on Emerging Threats and Capabilities of the Senate Committee on Armed Services, 106 Cong. 1 sess., April 20, 1999, p. 20; David A. Fulghum, "Unmanned Strike Next for Military," *Aviation Week and Space Technology*, June 2, 1997, pp. 47–48; Robert Holzer and Mark Walsh, "U.S. Navy Eyes Lethal UCAVs," *Defense News*, October 13–19, 1997, p. 1; David A. Fulghum, "JSF to Spawn Black Derivatives," *Aviation Week and Space Technology*, March 9, 1998, p. 55.

20. Steven Ashley, "Palm-Size Spy Planes," *Mechanical Engineering*, vol. 120 (February 1998), pp. 74–78; David A. Fulghum, "Miniature Air Vehicles Fly into Army's Future," *Aviation Week and Space Technology*, November 9, 1998, pp. 37–38.

21. Jack Hollingum, "Military Look to Flying Insect Robots," *Industrial Robot*, vol. 25 (1998), pp. 124–28.

22. David A. Fulghum, "Small Stealth Designs within China's Grasp," *Aviation Week and Space Technology*, June 7, 1999, pp. 28–29; Ivan Amato, "May the Micro Force Be with You," *Technology Review* (September–October 1999), pp. 74–82.

a decade after the first.[23] In other words, fuel efficiency and basic rocket design have been improving very modestly. Most boosters in use today are similar or identical to those in use more than a decade ago. Indeed, the last major U.S. effort to develop a new rocket engine was in the space shuttle program two decades ago.[24] In addition, the theoretical maximum performance of current chemical fuels is being approached, leaving increases in rocket size or modest reductions in body weight as the most practical—but inherently limited—means of improving booster performance.[25] Nor has the reliability of rockets—the best of which historically have approached roughly a 98 percent successful launch rate—improved of late. Indeed, in 1998 and 1999 the U.S. launch industry had one of its worst stretches in decades, with a total of six major failures—half with military payloads, half with civilian—over a nine-month period.[26]

It is possible that the next twenty years will see more breakthroughs than the last twenty. However, the slow evolution of rocket technology in recent years should give pause to highly optimistic projections. Notably, the next-generation U.S. rocket system, the so-called evolved expendable launch vehicle program, is not expected to reduce costs by even half. And it may very well be the last new launch vehicle until after 2020.[27]

Some other possibilities exist, such as the so-called rotary rocket in which centrifugal force from a rapidly spinning engine block rather than large turbopumps would provide the pressure for combustion. If such an idea proves feasible, it could lower launch costs by a larger amount. It could also reduce the odds of catastrophic launch failure, since many of its systems could be fired up and tested while the vehicle was still on the ground. Nevertheless, the proposal is seen as very challenging.[28]

23. Timothy M. Laur and Steven L. Llanso, *Encyclopedia of Modern U.S. Military Weapons* (Berkley Books, 1995), pp. 87–96, 466–68.

24. Craig Covault, "Rocket Propulsion Tests Aimed at X-33, X-34 and Delta 4," *Aviation Week and Space Technology*, May 4, 1998, pp. 51–53.

25. Office of Technology Assessment, *Launch Options for the Future* (U.S. Congress, 1988), pp. 19–24; Curtis D. Cochran, Dennis M. Gorman, and Joseph D. Dumoulin, eds., *Space Handbook* (Maxwell Air Force Base, Ala.: Air University Press, 1985), pp. 3-21 through 3-35.

26. Kathy Sawyer, "Rocket Failures Shake Faith in Space Industry," *Washington Post*, May 11, 1999, p. A1.

27. Ira F. Kuhn Jr., "Potential for Long Standoff, Low Cost, Precision Attack," in Defense Science Board 1996 Summer Study Task Force, *Tactics and Technology for 21st Century Military Superiority*, vol. 3, *Technology White Papers* (Department of Defense, 1996), p. III- 28.

28. Tim Beardsley, "The Way to Go in Space," *Scientific American*, February 1999, p. 85; "On a Rotor-Blade and a Prayer," *Economist*, March 20, 1999, pp. 83–84.

Ramjet and scramjet engines also offer potential for long-term advances in rocketry. These could be useful for both space launch and aircraft, as they combine the benefits of rockets (high speed) with those of jets (taking in oxygen from the air rather than having to carry their own supply).[29] Unlike normal jet engines, ramjets and scramjets do not require fan blades to compress and heat air so that fuel injected into it will burn; the aircraft's own speed provides the necessary compression. Ramjets take in supersonic air for the combustion process, but the air is subsonic by the time combustion occurs. In scramjets, if they prove practical, air within the combustion chamber will remain supersonic.[30]

If they prove workable, these propulsion systems could be at least two or three times as efficient as current rockets. By giving munitions greater speed and kinetic energy, they could also improve their ability to penetrate buried or hardened targets.[31]

Research efforts to design hypervelocity missiles and aircraft remain preliminary, however. The technical hurdles to success, not just in terms of achieving high speeds but also of maneuvering accurately at such speeds, are rather daunting.[32] Hans Mark, the Pentagon's director of reseach and engineering in the late 1990s, considers hypersonic aircraft still to be far in the future. Caution is indeed called for in forecasting the arrival of hypersonic "air-breathing" technologies. After all, research has been conducted on them for decades—most recently in the National Aerospace Plane project of the 1980s and early 1990s—and technical progress has been slow and difficult. Even if the initial concepts work, it is doubtful that hypersonic vehicles will be inexpensive enough and carry large enough payloads to be

29. Charles R. McClinton, "Air-Breathing Engines," *Scientific American*, February 1999, p. 84.

30. Michael A. Dornheim, "Missiles Lead Hypersonics Revival," *Aviation Week and Space Technology*, October 13, 1997, pp. 62–65.

31. William B. Scott, "Space Access' Launch System Based On Airbreathing Ejector Ramjet," *Aviation Week and Space Technology*, March 30, 1998, pp. 75–77; Robert Wall and David A. Fulghum, "Combat Weaknesses Trigger New Research," *Aviation Week and Space Technology*, February 16, 1998, p. 25; Grant R. Fowles, *Analytical Mechanics*, 3d ed. (Holt, Rinehart and Winston, 1977), p. 183; Cochran, Gorman, and Dumoulin, *Space Handbook*, pp. 3-1 through 3-7; Covault, "Rocket Propulsion Tests Aimed at X-33, X-34 and Delta 4," p. 52; T. K. Mattingly, "A Simpler Ride into Space," *Scientific American*, October 1997, pp. 120–25.

32. William B. Scott, "Lockheed Martin EELV Designed for Flexibility," *Aviation Week and Space Technology*, July 20, 1998, pp. 56–57; Joseph C. Anselmo, "EELV Win Boosts Boeing Launch Plans," *Aviation Week and Space Technology*, October 26, 1998, p. 71; David A. Fulghum, "Mach 8 Missiles Need More Work," *Aviation Week and Space Technology*, August 24, 1998, p. 24; Scott Gourley, "Soaring Ambitions," *Jane's Defence Weekly*, October 21, 1998, pp. 36–37.

used in anything besides very special-purposes missions.[33] One concept called Hypersoar has been developed at Lawrence Livermore National Laboratory and might prove capable of carrying substantial payloads, but it remains in a preliminary state of conceptual development.[34]

Ships

In recent decades, the basic propulsion systems, shapes, and structures of ships have changed only modestly. Some new capabilities have been developed, including landing-craft air cushion (LCAC) vessels designed to minimize friction with water and permit higher-speed transit. However, major carrier and surface combatant vessels retain sizes, speeds, and power plants similar to those they have fielded for many years (see table 4-3).[35] They continue to be vessels traveling at perhaps thirty to thirty-five knots, with large radar cross sections of several hundred square meters; wakes that can be observed for tens of kilometers; limited resilience against small numbers of hits by missiles; and limited maneuverability.[36]

What about the future? Progress is likely on a number of fronts, some of it impressive. It might well prove feasible to reduce ship radar cross sections, as well as other types of signatures, by 50 to 75 percent or more.[37] Electric-drive propulsion would allow a single power system for the entire ship (rather than one for electricity and one for movement). It could also make possible improvements in ship propulsion and design by permitting motors to be distributed in various places in the ship. It could also make submarines quieter, improving their survivability.[38] Water-jet propulsion and a ship design that

33. John A. Tirpak, "Mission to Mach 5," *Air Force Magazine,* vol. 82 (January 1999), pp. 28–33; Clark and Mulholland, "Pentagon Science Chief Pushes Leap-Ahead Research," p. 6.

34. "London to San Francisco in Less than 2 Hours," *Science and Technology Review* (December 1998), p. 2.

35. This can be seen by comparing, for example, classes of major vessels built in the 1960s and thereabouts to those built in the 1980s; typically, tonnage, speed, range, and power vary by no more than 10 to 20 percent. See Captain Richard Sharpe, ed., *Jane's Fighting Ships 1995–96,* 98th ed. (Surrey, England: Jane's Information Group, 1995), pp. 801–17.

36. In terms of maneuverability, ships today generally have turning radii of no more than two degrees per second (meaning it takes nearly a minute to change direction from north to east, for example). See Erbil Serter, "Warship Designs for the 21st Century," *Jane's International Defense Review,* quarterly report no. 4 (1997), p. 3.

37. See David Foxwell and Joris Janssen Lok, "Approaching Vanishing Point: The Emergence of Stealth Ships," *Jane's International Defense Review,* vol. 31 (September 1998), pp. 43–48; Ben R. Rich and Leo Janos, *Skunk Works* (Boston: Little, Brown, 1994), pp. 271–78.

38. Defense Science Board Task Force, *Submarine of the Future,* briefing charts, Office of the Under Secretary of Defense for Acquisition and Technology, 1998.

Table 4-3. *Maximum Speed of Destroyers over Time*

Hull number	Class designation	Initial operational capability	Speed (knots)
DD-1	Bainbridge	1902	29
DD-17	Smith	1909	31
DD-51	O'Brien	1915	29
DD-75	Wickes	1918	36
DD-186	Clemson	1919	36
DD-348	Farragut	1934	37
DD-356	Porter	1936	35
DD-421	Benson	1940	37
DD-445	Fletcher	1942	36
DD-710	Gearing	1945	35
DD-931	Forrest Sherman	1955	30
DDG-2	Charles F. Adams	1960	30
DDG-40	Coontz	1960	35
DD-963	Spruance	1975	33
DDG-51	Arleigh Burke	1989	30

Source: "Directory of American Naval Fighting Ships: 1996–99," U.S. Naval and Shipbuilding Museum (www.uss-salem.org/danfs/destroy/ [November 1997]).

causes less drag on the vessel from its captive waves may make it possible for cargo ships to roughly double their speeds to about forty knots.[39] Catamaran hulls and other innovations may allow speeds approaching fifty knots, at least for certain types of large ships, and speeds near thirty knots for small vessels.[40] Larger propellers could reduce the acoustic signatures of ships by raising the speed at which cavitation—or the production of small bubbles, and hence noise, by the ship's propellers—begins (in current ships this process begins at speeds of roughly twenty knots).[41] Reinforced hulls are likely to be able to increase survivability, particularly on large sea vessels with no need for great speed such as a mobile sea-going base or an arsenal ship holding hundreds of missiles and lurking low in the water.[42]

Despite these improvements, ships will remain slow, visible, and vulnerable relative to most other weapons platforms. They will continue to take

39. David L. Giles, "Faster Ships for the Future," *Scientific American*, October 1997, pp. 126–31.
40. David Mulholland, "New Vessel Promises High Speeds in Rough Seas," *Defense News*, December 7, 1998, p. 36.
41. Jacquelyn K. Davis, *A Smart Carrier for the New Era* (Washington, D.C.: Brassey's, 1998), p. 63.
42. Serter, "Warship Designs for the 21st Century," pp. 5–15; "U.S. Navy Investigates Sea-Going Base," *Aviation Week and Space Technology*, July 7, 1997, p. 68; Glenn W. Goodman Jr., "Redesigning the Navy," *Armed Forces Journal International*, vol. 135 (March 1998), pp. 28–33; Lucy Howard, "Faster Troop Carriers," *Newsweek*, May 25, 1998.

a week or weeks to cross major oceans; remain by far the most visible objects to radar and most other sensors on the ocean surfaces; and be valuable enough to justify many shots from an adversary, even if the probability of any one of those shots sinking the vessel declines.

Ground Vehicles and Armor

Most modern tanks, whether Western or Russian, weigh fifty to seventy tons, can attain speeds of forty miles per hour, possess night-vision capabilities and laser range finders, and have stabilized turrets so as to fire accurately while on the move. Their armor is commonly either composite—including metals like aluminum and perhaps even some uranium, glass, and fibers—or reactive, meaning that elements of it detonate when struck so as to drive away shaped charges.[43] These types of features made the M1 Abrams remarkably lethal, and survivable, in Desert Storm. For example, in at least one engagement 85 percent of all shots struck their targets.[44] No Abrams were permanently lost due to enemy fire in the war.[45]

But antiarmor weapons are getting better, too, meaning that, on balance, tanks have had to become even heavier to be reasonably survivable. Virtually all remain vulnerable, even to antitank rockets if those rockets are fired from the right angle and a close range. The latest portable antitank weapons are designed to attack through a tank's lighter top armor, as are the homing or autonomous submunitions fired by artillery such as the sense and destroy armor munition (SADARM).[46] Even as tanks keep getting better,

43. Laur and Llanso, *Encyclopedia of Modern U.S. Military Weapons*, pp. 228–35; James K. Morningstar, "Technologies, Doctrine, and Organization," *Joint Forces Quarterly*, no. 15 (Spring 1997), p. 39.

44. See Malcolm Chalmers and Lutz Unterseher, "Is There a Tank Gap? Comparing NATO and Warsaw Pact Tank Fleets," *International Security*, vol. 13 (Summer 1988), pp. 38–42; Steven J. Zaloga, Malcolm Chalmers, and Lutz Unterseher, "The Tank Gap Data Flap," correspondence, *International Security*, vol. 13 (Spring 1989), pp. 180–94; Stephen Biddle, "Victory Misunderstood: What the Gulf War Tells Us about the Future of Conflict," *International Security*, vol. 21 (Fall 1996), pp. 153–65; "Syria Deploys Upgraded T-55s near Golan Heights," *Jane's Defence Weekly*, August 20, 1997, p. 3; Robin Fletcher, "Dooming the Behemoths," *Armed Forces Journal International*, vol. 134 (May 1997), pp. 35–43.

45. Department of Defense, *Conduct of the Persian Gulf War: Final Report to Congress* (April 1992), p. T-145.

46. Institute of Land Warfare, "The Division Advanced Warfighting Experiment: Fire Support Implications (Crusader, MLRS and Comanche)," Defense Report 98-3 (Arlington, Va.: Association of the U.S. Army, April 1998).

overall trends appear to favor the weapons shooting at them, which have been proving lethal at increasingly long ranges for decades.[47]

Partly in recognition of the modest pace of technical improvement in armor and vehicle design, no Western country has firm plans to field a new battle tank before 2015 or perhaps even 2020. In the aftermath of the 1999 war against Serbia, when it was criticized for being too "heavy," the army has begun to place more rhetorical emphasis on a new and lighter tank, but it may not be able to build it until 2020 or so. Improvements until then, as with the army's M1A2 upgrade to the Abrams tank, will focus largely on night-vision, communications, and computing capabilities as well as protection against weapons of mass destruction. They will also provide tanks with longer-range munitions fired from their main guns, perhaps doubling the typical maximum range of three to four kilometers found today. Some of these munitions may have the capability to follow a laser designator or other terminal-homing mechanism to target.[48]

Even next-generation systems may not incorporate radically new technology. For example, advanced composite materials will do well to become twice as protective as modern steels, per unit of weight, between now and the year 2020.[49] There are aspirations for greater weight reductions from institutions such as the Army Research Laboratory, but few specifics about the nature of the purported super-light materials or the time horizon over which they might be developed have been provided.[50] Even if progress in armor and engine technology accelerates, moreover, the improving capabilities of antitank weapons may impede progress toward a future tank design less than half as heavy as today's.[51]

One might elect to build a lighter tank even without technological breakthroughs, but that would reflect a calculated gamble that the relative benefit of agility and deployability was worth a reduction in survivability. The

47. Stephen Biddle, "The Past as Prologue: Assessing Theories of Future Warfare," *Security Studies*, vol. 8 (Autumn 1998), pp. 1–74.

48. Daniel G. Dupont, "Pressured to Define Future Combat Vehicle, Army Asks DARPA for Help," *Inside the Pentagon*, July 1, 1999, p. 1; Christopher F. Foss, "Modernisation Points Way to Future MBTs," *Jane's Defence Weekly*, June 18, 1997, pp. 23–26; Rupert Pengelley, "Focusing Firepower for the Future," *Jane's International Defense Review*, vol. 31 (June 1998), pp. 44–52; Jason Sherman, "Souped up Big Time," *Armed Forces Journal International*, vol. 135 (May 1998), pp. 42–47.

49. National Research Council, *Strategic Technologies for the Army of the Twenty-First Century: Airborne Systems* (Washington, D.C.: National Academy Press, 1993), p. 43.

50. Dupont, "Pressured to Define Future Combat Vehicle," p. 1.

51. Vago Muradian, "Mark: U.S. Needs 20-Year View to Develop Its Future Weapons," *Defense Daily*, November 5, 1998, p. 1.

army's armored gun system, canceled in 1996, was to be just such a tank and weigh only about one-third as much as the Abrams. Such a weapon makes sense as an added tool for a military that may face relatively weak adversaries in distant locales, but that does not mean it—or anything else like it—will constitute a truly radical capability anytime soon.[52]

It is possible that composite-material vehicles lacking radar-reflective metal may become practical by 2020. Designed to be stealthy, such "plastic" vehicles might also carry a wide array of countermeasures against incoming precision munitions. Their viability hinges, however, on breakthroughs in a wide range of supporting technologies, including electric motors and guns.[53]

As noted, new ideas in gun technology are being explored; they include propelling projectiles by igniting plasmas or passing them through electromagnetic fields. If these ideas work, the energy content of future tank rounds—the best measure of penetration capabilities—may increase by perhaps 50 percent. However, they may not prove practical. For one thing, the energy storage per unit weight of capacitors is currently a factor of ten less than what would be needed to make this approach successful. Current guns can fire only a few rounds with the mobile power sources now available for them; in fact, they wear themselves out after that number of shots as well, requiring major repairs thereafter.[54] (Electric guns may be more practical for firing small rounds; an experimental handheld version can already fire hundreds of times more rounds per second than current Gatling or machine guns, though at present the rounds are quite small in size. If it advances, this type of technology could prove useful for clearing minefields or shooting down incoming missiles, whether or not related technologies prove viable for tank guns.)[55]

In addition to the incrementalism that characterizes advances in armor, propulsion and transmission systems are likely to remain similar to their current variety. Internal-combustion and gas turbine engines may well remain the prevalent means of propulsion for ground vehicles for several decades and will do well to achieve a doubling of fuel efficiency. That would be impressive, but hardly enough, for example, to fundamentally change the

52. Sean D. Naylor, "Welcome Mat Pulled from Combat System," *Army Times*, April 20, 1998.

53. Hugh McManners, "Plastic Tank Is Silent Killer of Battlefield," *Sunday Times* (London), February 7, 1999.

54. R. M. Ogorkiewicz, "In Search of Lighter, Smaller Electric Guns for Future Tanks," *Jane's International Defense Review*, vol. 32 (February 1999), pp. 26–31.

55. Dan Drollette, "Taking Ballistics by Storm," *Scientific American*, April 1999, p. 50.

need for huge fuel and logistics pipelines to support deployed forces.[56] As one revealing example of the military's own expectations about progress in ground vehicle technology, the army has recently approved plans to keep its 100,000 or so HUMVEEs through 2020 using a program of modest upgrades and refurbishments. That means that cold war–era vehicle technologies will continue to be found in large quantities on the battlefield of 2020.[57]

Battery-powered devices will not solve the problem of having to transport huge amounts of fuel overseas. At present, gasoline stores more than 100 times as much energy per unit of weight as even the best batteries, and the situation is unlikely to improve much in the batteries' favor in the near future. To put the situation starkly, a battery large enough to power a car some 500 miles would weigh more than the car itself and take up as much space as the passenger compartment.[58] If battery-driven battlefield vehicles ever became practical, the benefits could be substantial, since fuel represents a hefty percentage of the total weight of supplies that must be transported to the battlefield for any extensive operation. (Petroleum products constituted about two-thirds of all supplies shipped in Operation Desert Storm.)[59] But that day is a long way off, quite likely well beyond 2020.

Hydrogen-fuel-cell vehicles may be a more promising way to reduce fuel transport needs.[60] In theory at least, the hydrogen could be produced out of water in the theater of operations through photovoltaic or other means. The technology is not yet efficient or compact enough for large-scale applications, but it is progressing.[61] (More promising is the idea of producing hydrogen from petroleum, which would not fundamentally change transportation requirements since petroleum would need to be found on site or brought in from a distance—just as is the case today.) Even if it were possible to make hydrogen fuel from water in large amounts, it would still be necessary to transport the energy-producing apparatus, be it a set of giant solar arrays or even small nuclear reactors on ships. That would increase

---

56. See Dieter Zetsche, "The Automobile: Clean and Customized," *Scientific American*, September 1995, p. 105.

57. Daniel G. Dupont, "Army Outlines New Acquisition Plan for DoD's Massive Fleet of HUMVEEs," *Inside the Pentagon*, January 14, 1999, p. 1.

58. Gary L. Hunt, "The Great Battery Search," *IEEE Spectrum*, vol. 35 (November 1998), pp. 21–28.

59. Department of Defense, *Conduct of the Persian Gulf War*, pp. E-29, F-26.

60. Tom Gilchrist, "Fuel Cells to the Fore," *IEEE Spectrum*, vol. 35 (November 1998), pp. 35–40.

61. David Mulholland, "Powering the Future Military," *Defense News*, March 8, 1999, p. 1.

the number of soft, critical targets to be defended in the rear logistical area at a time when most agree the United States must try to decrease such vulnerabilities. It would also depend, obviously, on being located near a large source of water. In addition, it would probably not obviate the need to carry an initial supply of fuel for the days before localized hydrogen production would have caught up with demand.

Unmanned tanks are a possibility in the twenty-year time frame considered here. Unmanned military vehicles for mine detection are already being developed, and the trend is sure to continue.[62] Remote-control guns and mortars are also being developed and seem likely to enter service sooner than robotic tanks.[63] However, even if unmanned tanks became practical, they would still require armor and large engines that would keep them reasonably heavy.[64]

Munitions and Submunitions

This category of military technology is quite broad, including weapons that attack various ground targets and aircraft and range from small submunitions designed to hit individual vehicles or even individual soldiers to large aircraft-delivered bombs and missiles weighing one to two tons apiece.

Perhaps the most impressive developments of recent decades have involved the maneuvering of guided missiles. Consider the evolution of short-range infrared-guided air-to-air dogfight missiles since 1960. Back then, missiles could turn at only ten g's or slightly more, and a pilot needed to launch the missile from virtually directly behind the enemy aircraft to be able to take it down. By the 1970s, accelerations were over twenty g's, and pilots could shoot from ten degrees off the line the missile would need to follow. By the 1990s, maneuverability of the best missiles exceeded sixty-five g's, and some could, in effect, be shot sideways at a target.[65] Surface-to-air

62. David Pugliese, "Canada Military, Industry to Build Mine Detector Robot," *Defense News*, June 21, 1999, p. 26.

63. George I. Seffers, "U.S. Army to Test TRAP Remote Control Gun Platform," *Defense News*, June 29, 1998, p. 28; Joris Janssen Lok, "Remote U.S. Mortar to Undergo Initial Tests," *Jane's Defence Weekly*, October 15, 1997, p. 25.

64. "U.S. Army Launches Study to Develop Unmanned Tanks," *Jane's Defence Weekly*, August 5, 1998, p. 19.

65. Mark Farrer, "A New Arms Race—Fourth Generation Short Range Air-to-Air Missiles," *Asia-Pacific Defence Reporter* (December 1997–January 1998), pp. 22–23; A. P. O'Leary, *Jane's Electro-Optic Systems 1997–98*, 3d ed. (Surrey, England: Jane's Information Group, 1997), pp. 394–98.

missiles are continually improving, too. The best can also accelerate at many tens of g's.[66]

Likewise, antiship missiles have become quite impressive. Perhaps the lead example is the French Exocet, a subsonic missile that skims the surface of the sea to evade detection. Its range, when fired from a ship, is some seventy kilometers; it operates on inertial guidance initially but homes in on its target with active (that is, completely autonomous) radar. It can also be fired from aircraft. The Exocet was used in combat by Argentina in the 1982 Falklands War (destroying two British ships) and was employed in the 1987 *Stark* tragedy in which an Iraqi aircraft struck an American ship with two Exocets, killing thirty-seven crew members.[67]

Missiles relying on radio emissions—notably, global positioning system (GPS) signals—and radar for guidance are quite good, though not as accurate as laser-, infrared-, or TV-guided bombs. The former include cruise missiles and joint direct attack munitions (JDAMs); their inaccuracies are typically ten to fifteen meters. However, when dropped by a B-2, or perhaps someday other aircraft equipped with equally good radars to those of the B-2 force, the inaccuracy of JDAMs can be halved; future cruise missiles may have miss distances of only five or six meters as well. By contrast, the latter weapons—using laser, TV, and infrared guidance—have typical miss distances of three meters or less.[68]

Antiradar missiles like the U.S. high-speed antiradiation missile (HARM) have been very effective in conflicts like Desert Storm against radars that signal their location and then keep emitting while the missile homes in on the inadvertent beacon. Radar operators have become more sophisticated as a result, shutting off their power frequently to throw off any antiradiation missiles that may be headed at them. However, the latest version of the HARM includes a GPS-based guidance system that allows a missile to maintain its trajectory toward a target even if the target's outgoing radar signal is interrupted.[69]

66. David Mulholland, "U.S. Army May Begin HumRAAM Acquisition in 2000," *Defense News*, July 13, 1998, p. 10; Barton Wright, *World Weapon Database*, vol. 1, *Soviet Missiles* (Lexington, Mass.: D.C. Heath and Company, 1986), p. 603.

67. Sharpe, *Jane's Fighting Ships 1995–96*, p. 216; Navy League of the United States, *The Almanac of Seapower 1994*, vol. 37 (January 1994), p. 138; Lon Nordeen Jr., *Air Warfare in the Missile Age* (Washington, D.C.: Smithsonian Institution Press, 1985), pp. 201–03.

68. David Mosher, *Options for Enhancing the Bomber Force* (Congressional Budget Office, July 1995), pp. 24–25; William B. Scott, "Bad Weather No Deterrent for New Long-Range Weapons," *Aviation Week and Space Technology*, May 3, 1999, pp. 66–67.

69. David A. Fulghum, "Prowlers Exploiting Foes' Communications," *Aviation Week and Space Technology*, March 15, 1999, pp. 52–54.

An antiradiation missile of a different sort might be developed for air-to-air purposes to home in on the radio and radar emissions of a target aircraft. It would share the advantages of any passive missile in that it would be hard for the target aircraft to see it coming. Reportedly such a missile may already be in production by Russia and Ukraine.[70]

Shorter-range air-to-ground missiles performed well in Desert Storm and Operation Allied Force, though in the anti-armor mode they required good weather and a release point close to their target. This dependency on good weather can be reduced if either forces on the ground or UAVs have laser designators, making the weapons effective as long as cloud cover and fog are not found at very low altitudes. In such situations, bombs can be dropped by pilots who cannot themselves see the ground or laser-designate, with the bombs subsequently picking up a reflected laser signal once they descend below a certain altitude. In some cases, however, it may be deemed too dangerous to send ground forces near enemy formations to designate. And UAVs may not have good enough sensors to allow positive identification of targets, rendering their designation capabilities moot. Such was generally the case in Operation Allied Force, particularly given NATO's scrupulous rules of engagement designed to prevent accidental killings of civilians. Over time, however, this limitation should be overcome.[71]

Next-generation munitions are even less constrained, requiring neither ground forces nor a nearby UAV. They have the ability to home on targets autonomously if delivered to the right general vicinity first. For example, the skeet submunition, just recently having entered production, has been impressive. Forty skeet are released from the air force's sensor-fuzed weapon (SFW); for every sensor-fuzed weapon fired at test ranges, an average of four vehicles have been hit, putting about half of them at least temporarily out of action. Its explosively formed penetrator is released well above a vehicle and hence is unlikely to be defeated by explosive/reactive armor.[72]

Not all challenges with these kinds of weapons have been solved, however. The SFW has had problems when dropped from high altitudes or when trying to find targets amid countermeasures. It is being upgraded and its performance is likely to improve, but enemy countermeasures may improve as well. Similar problems afflict the army tactical missile system (ATACMS)

70. Douglas Barrie and Simon Saradzyhan, "Ukraine, Russia Clash over Marketing of New Missile," *Defense News*, May 31, 1999, p. 10.

71. David A. Fulghum, "Kosovo Conflict Spurred New Airborne Technology Use," *Aviation Week and Space Technology*, August 23, 1999, p. 30.

72. David A. Ochmanek and others, *To Find, and Not to Yield* (Santa Monica, Calif.: RAND, 1998), pp. 19, 36, 41–42, 93–94.

and associated submunitions.[73] Performance of munitions such as the SFW and brilliant antitank weapon (BAT) tends to be degraded by a factor of five or so in complex terrain or in the face of countermeasures.[74] Whether these developments represent a major new advantage for antiarmor weapons, or whether tanks will deploy countermeasures and possibly heavier top armor that are largely successful in thwarting these new tank killers, remains to be seen.[75]

Another type of weapon that builds on existing technology is an autonomous vehicle with its own small engine and seeker that can find a target and then direct an explosively formed penetrator warhead to destroy it. The low-cost autonomous attack system (LOCAAS) does this, using a laser radar (lidar) seeker for high resolution and employing many commercial off-the-shelf electronics in the lidar as well as the navigation and control systems. The goal is to produce these autonomous systems at less than $50,000 a copy. Whether that price goal is met, and whether the seeker is determined to be dependable enough that LOCAAS can be allowed to identify and destroy objects without human confirmation that they are indeed proper military targets, remains to be seen.[76] But in some settings, this weapon will also provide impressive autonomous attack capability and, like the SFW, allow attacks even when cloud ceilings are moderately low.

Other advances will result simply from combining various munitions and submunitions in new permutations. For example, rather than focus on large fixed targets, the navy's Tomahawk cruise missile could carry submunitions to attack tank and other vehicle formations. It might, for example, be positioned over a battlefield at a key moment in a campaign, hovering there until enemy troops expose themselves and become vulnerable to attack from above.[77]

73. Director of Operational Testing and Evaluation, *Operational Test and Evaluation: Annual Report 1997* (Department of Defense, 1997).

74. Defense Science Board 1998 Summer Study Task Force, *Joint Operations Superiority in the 21st Century*, vol. 2, *Supporting Reports* (Department of Defense, 1998), p. 27.

75. Eric H. Biass, "Portable Anti-Armour Missiles. A Prelude to the Tanksong?" *Armada International*, vol. 12 (February–March 1997), pp. 12–24; Defense Advanced Research Projects Agency, *Technology Transition* (Department of Defense, 1997), p. 62 (available at (www.darpa.mil/acrobat/transition.pdf [September 1999]); Muradian, "Mark: U.S. Needs 20-Year View," p. 1.

76. Clifford Beal, "Redesign for LOCAAS Air Weapon," *Jane's Defence Weekly*, June 16, 1999.

77. Robert Holzer, "Boosting Tomahawk's Punch," *Defense News*, July 12, 1999, p. 1.

Kinetic Energy Weapons

Super-high-velocity projectiles fired from guns may become possible in the next two decades. Such projectiles could be intercontinental in range and allow half-hour response times to points anywhere on the globe without nearly the cost of firing a ballistic missile. They would depend on innovations in gun technology that would permit roughly a tenfold increase in projectile speed and sufficient strength in the gun barrel to accept such wear. New types of hydrogen gas guns and electromagnetic rail guns offer at least the theoretical potential to achieve these speeds, as discussed previously in regard to tank technology, but they are a long way from being realized.[78]

Others have proposed using ballistic missiles to launch conventional warheads or metal rods that would destroy targets with their kinetic energy. Alternatively, the rods could be placed in orbit, being directed downward at tremendous speed on short notice. Some have claimed that hypervelocity rods could have accuracies of several meters using GPS guidance.[79] But that seems very optimistic given how little is known about the reentry of such rods. It seems sounder to assume accuracies in the modern intercontinental ballistic missile/submarine-launched ballistic missile (ICBM/SLBM) class of roughly 100 meters or somewhat better, meaning that the rods would be of little or no use against moving targets.[80] In that case, their tremendous speed would lose a good deal of its practical benefit.

The technical hurdles to making any such systems work are considerable. Even if they prove possible, they could be dangerous if mistaken for incoming nuclear weapons or if seen as a major threat to key hard targets like command and control facilities or ICBM silos. Nonetheless, they would possess noteworthy attributes. They would be able to penetrate air defenses reliably, and, if they carried many dozens of tungsten rods, one or two might hit a small target simply by chance.[81]

78. Harlan K. Ullman and James P. Wade Jr., "Rapid Dominance: A Force for All Seasons," paper prepared for the 1998 Joint Operations Symposium, National Defense University, September 9–10, 1998, pp. 46–49.

79. See Defense Science Board 1998, *Joint Operations Superiority in the 21st Century,* vol. 2, pp. 32–34.

80. Daniel G. Dupont, "Army Space Planners See Role in Urban Operations," *Inside the Pentagon,* February 18, 1999, p. 11.

81. Defense Science Board 1998, *Joint Operations Superiority in the 21st Century,* vol. 2, pp. 2–5.

Explosives

Current explosive technology allows standard bombs of roughly 1,000 to 2,000 pounds to penetrate at least several feet of concrete, even at relatively modest impact speeds of roughly the speed of sound (in the vicinity of 300 meters per second).[82] Recently, a more advanced weapon known as the GBU-28 has been developed. Its sophisticated sensors and fusing, together with the high speed of the incoming projectile and a total weight of 5,000 pounds, allow it to penetrate up to several floors of an underground structure before sensing that it has reached its target and detonating.[83]

However, kinematics and chemistry limit the further potential of weapons based on these principles. Conventional warheads are not becoming radically more powerful per unit of weight. Improvements in the basic chemical processes that give rise to their explosive power are expected to be modest at best in the future—perhaps 25 to 50 percent, but probably not much more. Even a doubling in power would increase the lethal radius of a given bomb against a given target by only 40 percent.[84]

Among other implications, these facts do not bode particularly well for efforts to destroy deep underground targets—where countries may choose to hide weapons or shelter military and political leaders and headquarters—with conventional munitions. Advances in drilling technology have been impressive. Current machines can dig a hole fifty feet in diameter and 200 feet in length in a single day, for example. That makes it relatively easy to put large numbers of bunkers many tens of feet below ground where they are hard to attack, even by weapons such as the GBU-28, particularly if they are also protected by thick layers of concrete.[85]

82. Mark Hewish, "Adding New Punch to Cruise Missiles," *Jane's International Defense Review*, vol. 31 (January 1998), p. 43; Tony Capaccio, "Tomahawk Strikes Again," *Washington Times*, February 19, 1998, p. B8.

83. David A. Fulghum, "Saudi Arabia Blocks USAF Warplane Shift," *Aviation Week and Space Technology*, February 16, 1998, pp. 22–26; Tim Smart, "Newest 'Smart' Weapons Would Redefine Air Campaign," *Washington Post*, February 1, 1998, p. A6; "Pentagon Tests 'Bunker Buster' for Iraq Raids," *Times* (London), January 30, 1998.

84. Kuhn, "Potential for Long Standoff, Low Cost, Precision Attack," p. III-28; Defense Science Board 1998, *Joint Operations Superiority in the 21st Century*, vol. 2, pp. 2–5.

85. Walter Pincus, "Buried Missile Labs Foil U.S. Satellites," *Washington Post*, July 29, 1998, p. A1.

Nonlethal Weapons

The technical underpinnings for a number of nonlethal warfare methods, such as computer virus attacks and radiofrequency weapons, have in some cases already been discussed in previous relevant sections. But there remain a number of other nonlethal technologies and techniques to consider.

Nonlethal weapons can be divided into a number of categories. One is that set of devices designed to control or otherwise affect individuals; a second focuses on vehicles and other motorized or electrical systems; and a third focuses on infrastructure like computers and communications.[86]

Today's military already deploys a number of nonlethal agents to affect people. These include pepper spray, wood batons, nonlethal wood and bean-bag rounds fired from shotguns, rubber bullets, sponge grenades, and sticky foam. Quasi-permanent dyes are also available to mark criminals, soldiers, or militia members so that they may be arrested once located.[87] Lasers that disable optics or temporarily blind or disorient enemy soldiers are also technically feasible, though legal limits have been placed on the use of those that could cause permanent injury.[88] Filaments to short out electricity-producing grids temporarily are available and have been used in recent operations, notably the 1999 air strikes against Belgrade.[89]

Among other interesting areas of nonlethal weapons research are those focusing on devices that might emit strong low-frequency acoustic waves with varying but severe effects on humans; "designer bacteria" that could destroy substances ranging from fuel to concrete or metal; disorienting strobe lights; acids to destroy the tires of vehicles; or airborne chemicals and fibers to wreak havoc with vehicle engines (see table 4-4).[90]

Not all of these weapons will become available by 2020; significant hurdles exist to making them operational. For example, a recent paper pre-

86. See Charles Heal, "Making, Not Breaking, the Rules," *Jane's International Defense Review*, vol. 30 (September 1997), pp. 77–80; Malcolm H. Wiener, *Non-Lethal Technologies* (New York: Council on Foreign Relations, 1995).

87. Heal, "Making, Not Breaking, the Rules," pp. 77–80; Mark Walsh, "Hurdles Remain for Deployment of U.S. Nonlethal Arms," *Defense News*, September 15, 1997, p. 14.

88. Robert J. Bunker, ed., *Nonlethal Weapons: Terms and References*, INSS Occasional Paper 15 (U.S. Air Force Academy, July 1997), p. 18.

89. Dana Priest, "NATO Hit Belgrade Grid with Classified, Lightning-like Missile," *Washington Post*, May 4, 1999, p. A15.

90. John B. Alexander, *Future War: Non-Lethal Weapons in Twenty-First-Century Warfare* (Thomas Dunne Books, 1999), pp. 60–80.

Table 4-4. *Demonstration and Validation Budget for Currently Funded Joint Nonlethal Weapons Programs, Fiscal Year 2000*
Dollars

Program	Budget
Modular crowd control munition (MCCM)	693,000
40mm nonlethal crowd dispersal munition	659,000
Nonlethal bounding munition	540,000
Canister-launched area denial system (CLADS)	1,674,000
UAV nonlethal payload/delivery system	108,000
66mm vehicle-launched munitions	1,911,000
Foam applications	1,017,000
Ground (electric) vehicle stopper	1,200,000
Vessel stopper system	1,200,000
Acoustic program	108,000

Sources: Department of Defense, *Joint Non-Lethal Weapons Program: 1998 A Year of Progress* (February 1999), pp. 5–10; Department of the Navy,"RDT&E Budget Item Justification Sheet (R-2 Exhibit)," Department of Defense, March 1999 (http://164.224.25.20/archives.nsf/sstatus/8525670e004c668a852566fe006ba4fc/$file/(u)75rdten_27.pdf [October 1999]).

sented at Lawrence Livermore National Laboratory included the following statement on acoustic weapons: "Substantial development work, both in terms of understanding the physiological effects and determining operational employment, will be required for this line of research to become useful."[91] There are other limits on nonlethal weapons as well. One important point is that the ranges of most nonlethal weapons are not, and are not expected to be, particularly great. For example, a laser device that would ionize air and hence allow passage of an electrical current to stun individuals might be useful out to 100 meters or so, if it can be successfully developed.[92]

Moreover, many nonlethal weapons will be indiscriminate-area weapons that would affect civilians and soldiers equally. That could place constraints on their use, particularly since depicting them as nonlethal may prove an exaggeration. Consider acoustic weapons again. To be truly effective, they may have to cause such intense effects that they will be potentially lethal, particularly to frail individuals or those exposed at close ranges.

A substance that greatly reduced traction and impeded flight operations could be used against runways, perhaps being dispersed by cruise missiles

91. Anthony Fainberg and Xavier Maruyama, "Overview of Key Technologies for Peace Operations," in Alex Gliksman, ed., *Meeting the Challenge of International Peace Operations: Assessing the Contribution of Technology* (Livermore, Calif.: Center for Global Security Research, Lawrence Livermore National Laboratory, 1998), p. 118.

92. David Mulholland, "Laser Device May Provide U.S. Military Nonlethal Option," *Defense News*, June 14, 1999.

or other precision-guided munitions. However, the types of mechanisms needed to deliver such nonlethal weapons could also deliver traditional ordnance, probably with no greater difficulty.

Thus nonlethal weapons are unlikely to become decisive instruments of high-intensity war. Nonetheless, they will increasingly become useful in situations where minimizing civilian or even enemy combatant casualties is an important constraint on military operations. There may be situations where disabling the runway, stunning an individual, or stopping a vehicle is the best recourse U.S. military forces have available to them.

Weapons of Mass Destruction

What of weapons of mass destruction? Beginning with nuclear weapons, the basic physics and technologies are unlikely to change appreciably. For example, possible development of a pure fusion weapon, perhaps triggered by a laser ignition system rather than a fission bomb primary, appears many decades away. Low-yield nuclear weapons that would detonate underground could provide a somewhat better weapon against hardened or deep underground targets than are now available and might destroy the target without causing extensive collateral damage to nearby civilian populations. However, it is doubtful that they would be highly usable weapons, since they would still be nuclear weapons. In any case, in light of the U.S. government's signature to the comprehensive test ban treaty, it is also unlikely they will be developed.[93]

As for the spread of nuclear weapons to other countries, a major technological concern is the possibility that gas centrifuges or laser-isotope devices may make "basement proliferation" possible in the future. Historically, producing nuclear materials has required large reactors or massive uranium enrichment processes based on gaseous diffusion principles. That has generally made it possible to find them by satellite. However, gas-centrifuge or laser-enrichment methods may not require such large facilities. Thankfully, they would require very specialized equipment, providing some hope that the efforts of would-be proliferators to import or build such equipment would be noticed.[94] Unfortunately, it may not be particularly dif-

93. See John Fleck, "Sandia Scraps Guided Nuclear Bomb Project," *Albuquerque Journal*, August 14, 1998, p. A1; "Destruction of Bunkers Is New Bomb's Function," *Washington Times*, April 14, 1997, p. A11.

94. David H. Albright, Barbara G. Levi, and Frank N. von Hippel, "Fissile Weapons Materials," in Frank von Hippel, *Citizen Scientist* (Simon and Schuster, 1991), pp. 105–16.

ficult to clandestinely build facilities large enough to produce enough material for just a few bombs.

Chemical weapons could have a wide range of effects depending on how proficiently they were dispersed and whether they were used as terrorist devices, against troops on the battlefield, or against U.S. military bases.[95] Indeed, had they been used in a battle like Desert Storm, they could have greatly complicated operations for U.S. forces, which would have had a hard time staying suited up for long periods in the desert environment and sustaining cleanup operations at the required pace. Casualties could have been numerous, and operations slowed considerably. Some of these vulnerabilities persist today.[96] Whatever the current situation, the most likely future trend appears to be a gradual improvement in protection and decontamination gear.[97] For example, enzymes that can consume chemical agents may be developed and replace the water-intensive techniques now in place. Protective suits and gear will improve, too.[98] Foams to contain and destroy agents dispersed within a confined space appear promising.[99]

Biological weapons, by contrast, may become intrinsically more dangerous with time. Under some circumstances biological weapons could already be as lethal as nuclear weapons—particularly if a country or group could disperse them aerially over a city without being detected and stopped, or if a truly horrific new agent with infectious qualities paralleling influenza or Ebola were developed and weaponized.[100] Biological weapons are even

95. Paul Mann, "Officials Grapple with 'Undeterrable' Terrorism," *Aviation Week and Space Technology*, July 13, 1998, p. 68.

96. See General Accounting Office, *Chemical Warfare: Soldiers Inadequately Equipped and Trained to Conduct Chemical Operations*, GAO/NSIAD-91-197 (May 1991); Brad Roberts, "New Challenges and New Policy Priorities for the 1990s," in Brad Roberts, ed., *Biological Weapons: Weapons of the Future?* (Washington, D.C.: Center for Strategic and International Studies, 1993), p. 73; General Accounting Office, "Chemical and Biological Defense: Observations on DoD's Plans to Protect U.S. Forces," GAO/T-NSIAD-98-83 (March 17, 1998), pp. 1–3.

97. See Victor A. Utgoff, *The Challenge of Chemical Weapons: An American Perspective* (St. Martin's Press, 1991), pp. 172–81; Office of Technology Assessment, *Proliferation of Weapons of Mass Destruction: Assessing the Risks*, OTA-ISC-559 (August 1993), pp. 52–57.

98. Mark Hewish and Joris Janssen Lok, "Air Forces Face up to NBC Reality," *Jane's International Defense Review*, vol. 31 (May 1998), pp. 49–50.

99. Malcolm W. Browne, "Chemists Create Foam to Fight Nerve Gases," *New York Times*, March 16, 1999, p. F2.

100. It is not trivial to develop and utilize such a dispersal mechanism or the agents themselves; even with considerable financial resources and a number of Ph.D.-caliber biologists, for example, Aum Shinrikyo failed on both counts. See Milton Leitenberg, "Biological Weapons Arms Control," PRAC Paper 16 (University of Maryland at College Park, Center for International and Security Studies, 1996), pp. 54–56.

Biological weapons could be remarkably lethal, however, especially if antibiotics were not quickly administered to the afflicted (or if they were not effective for the agent at issue). See

harder to monitor and control than nuclear weapons, for a number of reasons. Natural infectious agents occur throughout the world, so the raw materials for biological weapons could never be completely monitored (even leaving aside the possibilities of genetic engineering). Biological production facilities can be much smaller and more numerous than small nuclear ones, as well as being inherently dual-purpose.

Some of the improvements in protective gear and decontamination technologies that are being developed against chemical weapons will also improve defenses against biological weapons. Decontaminants may allow biological agents to be killed with far less corrosive effect on equipment than is presently possible. Moreover, sensors to detect biological agents will improve, as noted in chapter 3. Overall, however, the threat will probably become more severe with time.[101]

On-site challenge inspections under a strengthened biological weapons convention could help. These may become possible without compromising commercially valuable proprietary information through new sampling techniques that do not require removing sensitive biological materials from the premises of the suspect site. This approach would probably catch some illicit sites. However, it would probably not find all such sites, and the inspections would not be useful in a country that declined to participate in the regime.[102] Making matters more complex yet, research to develop antidotes or protective measures against various potential weapons provides a legitimate pretext for working with agents that could be weaponized—so even a smoking gun obtained during a challenge inspection would not always provide conclusive proof of noncompliance.[103]

Office of Technology Assessment, *Proliferation of Weapons of Mass Destruction*, pp. 52–54; Mann, "Officials Grapple with 'Undeterrable' Terrorism," p. 67. See also John D. Steinbruner, "Biological Weapons: A Plague upon All Houses," *Foreign Policy*, no. 109 (Winter 1997–98), pp. 85–86; Office of Technology Assessment, *Technologies Underlying Weapons of Mass Destruction* (1993), pp. 93–117; Nicholas Wade, "Tests with Anthrax Raise Fears that American Vaccine Can Be Defeated," *New York Times*, March 26, 1998, p. A24; Ronald M. Atlas, "Medical/Biological Nature of the Threat of Biological Weapons to U.S. Security," paper presented at the Brookings Institution, April 27, 1998 (www.brook.edu/fp/events/19980427_atlas.htm [September 1999]).

101. Fernandez, statement before the Subcommittee on Emerging Threats and Capabilities, p. 6.

102. Thomas P. Monath and Lance K. Gordon, "Strengthening the Biological Weapons Convention," *Science*, November 20, 1998, p. 1423.

103. See Roberts, "New Challenges and New Policy Priorities for the 1990s," pp. 68–101; Judith Miller and William J. Broad, "Germ Weapons: In Soviet Past or in the New Russia's Future?" *New York Times*, December 28, 1998, p. A1; Atlas, "Medical/Biological Nature of the Threat of Biological Weapons to U.S. Security."

Missile Defenses

Missile defense proponents are right that the global threat is growing. The club of nations with ballistic missiles now numbers about two dozen.[104] Just as important is the increasing range of the missile forces of many countries. Notable in this regard is North Korea, which in the summer of 1998 conducted a test of a multistage missile, the Taepo Dong 1. It is also working on a Taepo Dong 2 that may have enough range to deliver a substantial payload against Hawaii or Alaska and perhaps even a nuclear payload against parts of the continental United States (assuming that North Korea can add a third stage to the rocket and build a reentry vehicle capable of surviving atmospheric reentry).[105]

The North Korean test confirmed the findings of a congressionally mandated commission led by former secretary of defense Donald Rumsfeld. The commission's report, published shortly before the Taepo Dong 1 overflew Japan and splashed down in the Pacific Ocean, concluded unanimously that one or more other countries might develop a missile threat against American territory fairly quickly. Criticizing the U.S. intelligence community, which had argued that we would have a decade or more of clear warning before most countries could acquire credible missile threats against the United States, it pointed out that we might only have five years' warning or perhaps even less. This is because would-be proliferators could do a good deal of their preliminary research in secret, conduct crash programs in missile testing (albeit at some price in missile capability and reliability), buy missiles from abroad, or find ways to threaten the United States with medium-range missiles launched from ships or the territories of nearby countries.[106]

It therefore makes sense to think hard about building ballistic missile defenses. But is that technologically realistic? After all, the Patriot missile performed relatively poorly in the Gulf War against a simple SCUD missile threat, by most independent accounts intercepting no more than 10 to 20 percent of the missiles against which it was fired.[107]

104. Dov S. Zakheim, "Old Rivalries, New Arsenals: Should the United States Worry?" *IEEE Spectrum*, vol. 36 (March 1999), pp. 30–31.
105. Harold A. Feiveson, ed., *The Nuclear Turning Point* (Brookings, 1999), pp. 68–69; Michael O'Hanlon, "Rethinking Star Wars," *Foreign Affairs*, vol. 78 (November–December 1999).
106. Donald H. Rumsfeld and others, "Executive Summary of the Report of the Commission to Assess the Ballistic Missile Threat to the United States," Washington, D.C., July 15, 1998, pp. 1–6, 11–13.
107. See Theodore A. Postol, "Patriot Experience in the Gulf War," correspondence, *International Security*, vol. 17 (Summer 1992), pp. 225–40. Postol not only presents results of his

Some progress is being made in short-range theater missile defense (TMD). For example, the United States has improved its Patriot theater defense system since the Gulf War. The Patriot PAC-2 system recently shot down four incoming missiles in four attempts in a training exercise. In addition, Patriot's radar should soon have a good ability to distinguish between heavier and lighter objects so that it will no longer be fooled by the breakup of a missile's body during atmospheric reentry (as it was in Desert Storm) or an enemy's use of simple decoys.[108]

Starting in 2001, the Pentagon is to deploy a further-improved version of the Patriot with a new hit-to-kill interceptor missile that achieved a completely successful test in early 1999. Whereas the existing Patriot system, known as PAC-2, can defend an area with a radius of some ten to fifteen kilometers, the new PAC-3 will triple that coverage. The interceptor has its own high-resolution radar and sophisticated computer processor aboard, as well as 180 small thrusters for fine steering in the final phases of approach to target (earlier Patriots have fins for steering as well as blast-fragmentation warheads).[109]

The Pentagon is also continuing to develop a low-altitude theater defense based on navy ships. Known simply as the navy area defense system, it has a coverage zone somewhat larger than that of the Patriot PAC-3. The navy hopes to deploy this system by 2003. Figure 4-3 shows how the navy area defense system is to complement Patriot and other TMD initiatives.

Less advanced are other programs designed to provide defense against theater missiles over regions of a few hundred kilometers' width. The key programs are known as theater high-altitude area defense (THAAD) and the navy theater-wide system (NTW). These have often been in the news because of their testing difficulties, though THAAD finally scored a direct hit during a test in June 1999 and another at slightly higher altitude and greater speed in August.[110] (The previous problems with THAAD tests had

---

own videotape analyses but quotes Israeli experts who reach similar conclusions and points out flaws in the official army analysis—such as a complete lack of chemical analysis of bomb craters to determine what produced them.

108. Daniel G. Dupont, "U.S., German, Dutch Troops Go Four-for-Four in Patriot Testing," *Inside the Pentagon*, July 1, 1999, p. 3; Bradley Graham, "Army Hit in New Mexico Test Said to Bode Well for Missile Defense," *Washington Post*, March 16, 1999, p. A7; James Glanz, "Missile Defense Rides Again," *Science*, April 16, 1999, p. 417; David Hughes, "Patriot PAC-3 Upgrade Aimed at Multiple Threats," *Aviation Week and Space Technology*, February 24, 1997, pp. 59–61.

109. Graham, "Army Hit in New Mexico Test Said to Bode Well for Missile Defense"; Glanz, "Missile Defense Rides Again."

110. The first test occurred below fifty miles' altitude; the second, above fifty miles. See "World News Roundup," *Aviation Week and Space Technology*, June 14, 1999, p. 56;

Figure 4-3. *Theater Missile Defense Architecture*

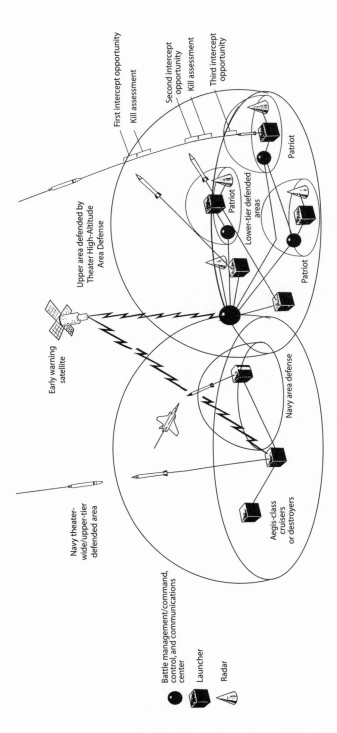

Source: David Mosher, "The Grand Plans," *IEEE Spectrum,* September 1997, pp. 28–29.

more to do with poor quality control and haste than fundamental immaturity of the new technologies involved. A faulty cable prevented proper separation of a booster from the rest of the rocket in one test failure; a battery connector shorted in another; the infrared sensor was contaminated and thus blinded in yet a third.)[111] Whichever of these programs turns out to advance more rapidly is to be fielded in 2007, and the other sometime thereafter, if the Pentagon keeps to its current plans. Some consideration has been given to accelerating the THAAD program in light of its recent testing successes. On the other hand, the Pentagon's director for operational test and evaluation, Philip Coyle, has emphasized that the summer 1999 THAAD tests were conducted under somewhat artificial circumstances and has advocated a path of continued caution for the near term.[112]

Of all the aforementioned TMD programs, NTW is the only one that raises any Anti-Ballistic Missile (ABM) Treaty compliance issues. The treaty permits all theater missile defenses without restriction but does not clearly define the demarcation point between theater and strategic missile defenses. In 1997 the United States and Russia reached an accord that defined as TMD any system using interceptors that do not exceed speeds of three kilometers per second and that are not tested against incoming warheads with speeds greater than five kilometers per second or ranges greater than 3,500 kilometers. The NTW program is to stay below the two latter thresholds, but its interceptor has a maximum speed greater than three kilometers per second, making its status somewhat ambiguous.[113]

However, advanced theater missile defense systems may raise ABM Treaty concerns on technical grounds, even if they are deemed compliant on legal grounds. If the United States develops a national missile defense (NMD) capability, and deploys the sensors and battlefield management capability needed to operate such a system, the technical possibility may exist that theater-defense missiles could be plugged into such a national missile defense

Robert Wall, "THAAD at Crossroads after Intercept," *Aviation Week and Space Technology*, August 9, 1999, pp. 29–31; Major General Peter Franklin, "DoD News Briefing," Ballistic Missile Defense Organization, August 19, 1999.

111. Bradley Graham, "Low-Tech Flaws Stall High-Altitude Defense," *Washington Post*, July 27, 1998, p. A1.

112. John Donnelly, "THAAD Intercepts Were Unrealistic, Top Tester Says," *Defense Week*, August 23, 1999, p. 1.

113. Lisbeth Gronlund, Letter to the Editor, *Arms Control Today*, vol. 28 (June–July 1998), p. 36; John Pike, "Ballistic Missile Defense: Is the U.S. 'Rushing to Failure'?" *Arms Control Today*, vol. 28 (April 1998), p. 10.

system.[114] That technical concern complicates the case for a national missile defense and may have to be seriously addressed in the process of renegotiating the ABM Treaty in the future. Specifically, the number of NTW interceptor missiles may need to be capped at a relatively modest level.

The Clinton administration intends to have an NMD system up and running by 2005 if possible. Its specific goals are to deploy about twenty very-high-speed exoatmospheric interceptors—and perhaps up to 100 eventually—at a single site, most likely in Alaska. Such a single site may be able to cover the entire fifty states against a long-range ICBM threat, though shorter-range missiles launched from ships or submarines could not be stopped—raising the possibility that the original single site might eventually be expanded to include other interceptor bases.[115]

Is this technologically realistic? Someday, hit-to-kill technology against an incoming missile flying a clear trajectory should work. 1999 was a good year for such technology, with successful tests of both the Patriot PAC-3 and THAAD, as well as a successful intercept by the prototype system for a national missile defense. Further progress can be expected. But the NMD program is still too hurried. Only three or four of twenty-one flight tests scheduled for its development program are to be completed by the spring of 2000, when a preliminary decision on deployment is to be made. That is the type of program schedule that made a 1998 task force headed by retired general Larry Welch argue that missile defense research programs were being pushed too rapidly in what amounted to a rush to failure. The Pentagon's Ballistic Missile Defense Organization acknowledges that there is no way a proper production-quality interceptor booster rocket can be fully developed and tested before 2003, so the decision in 2000 will at most start the ball rolling. It is unreasonable to expect NMD to be deployable before shorter-range systems like THAAD or NTW can be fielded. Yet, with neither THAAD nor NTW to be deployed before 2007, the Pentagon intends exactly that.[116]

An even greater challenge, and one that will not be so easily solved by a simple scheduling change, is to discriminate advanced countermeasures from actual warheads and then intercept the latter. This may not always be easy even within the atmosphere. For one thing, infrared-homing interceptors have a hard time working in the lower atmosphere (where air resistance

114. Feiveson, *The Nuclear Turning Point*, pp. 87–89.
115. Feiveson, *The Nuclear Turning Point*, pp. 73–74.
116. General Larry Welch and others, "Report of the Panel on Reducing Risk in Ballistic Missile Defense Flight Test Programs," Department of Defense, February 27, 1998, pp. 7–27; Jeffrey A. Merkley, "Trident II Missile Test Program," Staff Working Paper (Congressional Budget Office, February 1986), p. 7.

is greatest and hence most useful for discriminating warheads from decoys), since their infrared viewing windows heat up from friction. Moreover, reentry vehicles could be built to maneuver within the atmosphere to try to evade interceptors (though it is uncertain they will be able to outmaneuver interceptors).[117]

Whatever the challenges within the atmosphere, things are much harder outside it, where NMD systems now under development would have to work.[118] Prior to a warhead's reentry into earth's atmosphere, air resistance will not have had a chance to separate out the (generally) lighter decoys from the heavier warheads.[119] Even extremely light decoys would fly the same trajectory as real warheads, so speed could not be used to distinguish the real from the fake. To mimic the infrared heat signature of a real warhead, thereby fooling sensors that measure temperature, decoys could be equipped with small heat generators. To fool radars or imaging infrared sensors, warheads and decoys alike could be placed inside radar-reflective balloons that would make it impossible to see their interiors.[120] Such countermeasures would doom a national missile defense of the type now under consideration and development in the United States.

The news is not entirely bleak, however. It is not entirely trivial to develop good decoy technology, including the means to dispense decoys in space—particularly for countries unable to do much missile testing. After all, the superpowers did not develop multiple independently targetable reentry vehicle (MIRV) technology until they had had long-range missiles for a decade or so. (MIRV technology is more complex than decoy technology, but there are parallels.) Although countermeasures would be unnecessary for an attack using many bomblets filled with chemical or biological agents, such attacks would be less dangerous than those with nuclear warheads. Chemical weapons are intrinsically less lethal than nuclear or biological agents. Biological agents are most lethal when distributed over a wide area by a device like a crop duster. They must also be released at the proper altitude, and kept at the proper temperature, to be effective—two difficult requirements to meet when delivering the agents by ballistic missile warhead. They also

117. Feiveson, *The Nuclear Turning Point*, pp. 77, 83.

118. George N. Lewis and Theodore A. Postol, "Future Challenges to Ballistic Missile Defense," *IEEE Spectrum,* vol. 34 (September 1997), pp. 60–68.

119. See Welch and others, "Report of the Panel on Reducing Risk in Ballistic Missile Defense Flight Test Programs," p. 56; Elaine M. Grossman, "Rumsfeld Commission Member Sticks to Guns on Opposing Defenses," *Inside the Pentagon*, July 30, 1998, pp. 19–20.

120. David C. Wright, testimony on the technical readiness of national missile defenses before the Senate Committee on Foreign Relations, 106 Cong. 1 sess., May 4, 1999; and Richard L. Garwin, testimony before same committee on same date.

tend to be most effective when the attacked country does not realize it has been attacked with biological agents until victims begin to show symptoms of disease (since antibiotics and other treatments are generally less effective after several days' incubation). Ballistic missiles are therefore less than ideal means of delivering such agents.[121] All in all, a light nationwide defense based on exoatmospheric missiles will have serious limitations but may still provide some capability against the type of threat a North Korea or Iran could develop in the next decade or two.

What about defenses against cruise missiles? This problem is not as technically daunting. However, neither is cruise missile production. It does not require either of the two most complex ballistic missile technologies: reentry vehicles or sophisticated inertial guidance systems. (GPS signals—as well as the potential to use terminal guidance based on radar, infrared, or laser homing—provide simpler guidance options.) So the threat may be more widespread than that posed by ballistic missiles and particularly by long-range ballistic missiles. Only three countries had land-attack cruise missiles at the close of the 1990s. But the antiship missile is already found in seventy countries in the world; some 75,000 are believed to exist. A number of countries are likely to develop land-attack cruise missiles in the years ahead, using such antiship cruise missiles as a technological stepping stone.[122]

Indeed, the United States already expects relatively stealthy cruise missiles—with radar cross sections of perhaps 0.01 square meters or less, in contrast to first-generation cross sections at least ten times greater—to be in the hands of a number of countries by 2005 or so. Cruise missile defenses involving a network of AWACS and JSTARS aircraft as well as advanced medium-range air-to-air missiles (AMRAAMs) or lower-cost interceptors may be effective by 2010. But the technical challenges of constructing the necessary integrated surveillance grid against low-flying and low-signature cruise missiles are daunting. Even if defenses work, cruise missiles could be used in a saturation attack or otherwise penetrate terminal defenses under some conditions. And cruise missile are certainly dangerous. On the battlefield, they will have the accuracy to attack key point targets that rudimentary ballistic missiles will not.[123]

121. Defense Science Board 1998, *Joint Operations Superiority in the 21st Century*, pp. 97–100; Office of Technology Assessment, *Proliferation of Weapons of Mass Destruction*, p. 52.

122. Dennis M. Gormley, "Hedging against the Cruise-Missile Threat," *Survival*, vol. 40 (Spring 1998), pp. 92–111; General Accounting Office, *Cruise Missile Defense*, GAO/NSIAD-99-68 (March 1999), p. 4.

123. David A. Fulghum, "Secret Upgrades Target Stealthy Cruise Missiles," *Aviation Week and Space Technology*, August 24, 1998, pp. 22–23; Lane Pierrot, *A Look at Tomorrow's*

Finally, what about more exotic technologies for shooting down either cruise or ballistic missiles? Notable are high-powered lasers, be they airborne, based in space, or based on the ground (and possibly assisted by mirrors in space, depending on their mission). These types of aspirations have been around since the days of the Strategic Defense Initiative in the 1980s; are they any closer to being realistic now?

The short answer is yes, but only to an extent. High-powered lasers are more impressive today than in the 1980s. The air force, in its airborne laser program for theater missile defense, reached the program's desired power thresholds with a prototype system in 1998.[124] However, the Pentagon's Philip Coyle subsequently raised fundamental concerns about the project's feasibility, including doubts that the actual laser system will be light enough to be airborne, concerns that sensors may not be able to distinguish countermeasures from actual rockets in their boost phases, and doubts that sensors will be able to stay locked onto a target. The General Accounting Office has also pointed out that the air force tests involved only a prototype laser device and did not create an actual projection of laser energy outside the device (that is, there was no beam in the commonly understood sense of the word).[125] Such concerns have not stopped air force visionaries from postulating airborne lasers as well as space-based laser systems with the ability to attack a host of ground targets.[126] However, the country probably is decades away from such capabilities. And even if the weapons someday prove useful against today's adversaries, that may change in the future. Missiles and aircraft could be built with thick or reflective skins or could be designed to rotate, deploy shields, and otherwise protect themselves if attacked. Moreover, operating a powerful laser weapon through the turbulent atmosphere remains a herculean task.[127] In addition to the matter of atmospheric turbulence is the fact, confronted in chapter 3, that many wavelengths of electromagnetic radiation do not travel easily through clouds or water vapor.

Tactical Air Forces (Congressional Budget Office, January 1997), p. 77; Michael C. Sirak and Daniel G. Dupont, "Study Group Warns Pentagon U.S. Is Vulnerable to Cruise Missile Attack," *Inside the Pentagon*, December 24, 1998, pp. 13–14; David Mulholland, "DARPA Aims to Hit Targets on the Cheap: Contract Finalists to Be Chosen Soon," *Defense News*, June 28, 1999, p. 4.

124. Michael E. Ryan, "The United States Air Force Posture Statement 1999," testimony before the House Armed Services Committee, 106 Cong. 1 sess., February 1999, p. 18.

125. John Donnelly, "Basis for Pentagon Approval of Airborne Laser 'Questionable,'" *Defense Week*, March 15, 1999, p. 1.

126. Thomas D. Bell, "Weaponization of Space: Understanding Strategic and Technological Inevitabilities," Occasional Paper 6 (Maxwell Air Force Base, Ala.: Air War College, 1999), p. 19.

127. Director of Operational Testing and Evaluation, *Annual Report 1997*.

More likely by 2020 is that directed-energy weapons on the ground, in the atmosphere, or in space could be a threat to satellites—especially those unable to alter their predictable orbits or otherwise protect themselves. As noted earlier, even China is believed to have some rudimentary capabilities of this type as of 1999.[128]

Conclusion

One of the most notable themes that emerges from the previous overviews of military technologies is that means of propulsion, while improving and becoming more efficient, show few signs of breakthroughs. Nor are the speed and weight of most major means of transportation, or of battlefield armored vehicles, likely to change drastically. As such, current constraints on moving and supporting large numbers of combat forces may be eased in the future, but they will not fundamentally change in nature or disappear. The aspirations of *Joint Vision 2010* for dominant maneuver and focused logistics thus appear rather optimistic, and the third technical premise of the RMA movement—as articulated in Chapter 1—is likely to prove far too optimistic.

Joint Vision 2010's goal of full-dimensional protection also appears too optimistic. Missile defense technologies are improving, for example, but only gradually and hardly enough to eliminate this type of threat to U.S. forces. It may prove possible to reliably stop an attack on national territory by just a few simple enemy missiles, or to intercept some considerable percentage of battlefield missiles, but the RMA and *Joint Vision 2010* notion of dependable protection goes too far.

The goal of precision engagement set out in *Joint Vision 2010* may be more practical—largely because it is a less absolute and lofty phrase, but also because greater progress is likely within the realms of munitions and specialized weapons. Autonomous and homing munitions and submunitions will improve in absolute terms over the next two decades, not so much because of the power of their explosives or other kill devices but because of the increased accuracy with which they will be delivered under a wide range of weather conditions. Whether they improve as dramatically in relative

128. See Robert Wall, "Intelligence Lacking on Satellite Threats," *Aviation Week and Space Technology*, March 1, 1999, p. 54; John Tirman, ed., *The Fallacy of Star Wars* (Vintage Books, 1984), p. 225; David A. Fulghum, "New Priorities for Pentagon," *Aviation Week and Space Technology*, November 9, 1998, p. 36; and Bill Gertz, "Chinese Army Is Building Anti-Satellite Laser Weapons," *Washington Times*, November 3, 1998, p. A1.

terms remains to be seen. That is, enemies may learn how to reduce the effectiveness of these munitions substantially through the use of sophisticated countermeasures; this is particularly likely in combat settings in cities, forests, and other complex terrain affording camouflage and shelter. On balance, the potential of these new weapons is considerable, but it will be realized to varying degrees depending on the scenario and the opponent.

Major strides are likely in areas of nonlethal and biological weaponry—for the better and for the worse, respectively. Changes in the former are unlikely to radically redefine warfare, however, due to their typically short ranges. Changes in the latter are best evaluated in the context of overall trends in controlling weapons of mass destruction, which are considered in more detail in chapter 6. Suffice it to say, the news is not particularly good.

Futuristic weapons such as airborne or space-based lasers, intercontinental artillery, and space-based tungsten rods able to fire downward quickly and accurately are not beyond the realm of the possible for the 2020 time frame. Likewise, large blimplike transport aircraft, hypersonic planes, scramjet-powered rockets, and "plastic tanks" are conceivable as well. But they are far from givens, and, in each case, major technical challenges cast doubt on their prospects. Simply being able to imagine such systems, and make laboratory progress researching some of their requisite constituent parts, is not a guarantee of imminent technological realization. In fact, some of these ideas have been under consideration for years or decades already, generally with little to show for the effort. It makes sense to keep researching them—and at least a couple will undoubtedly pan out within two decades—but we are hardly in a position to base future military planning on them.

CHAPTER FIVE

Toward a Verdict on the RMA Hypothesis

$\mathbf{W}$hat does the survey of trends in key military technologies presented in chapters 3 and 4 and summarized below in box 5-1 portend for future warfare? And what does it suggest about the validity of the most common form of the hypothesis that a revolution in military affairs (RMA) is now within reach?

The current RMA hypothesis is summarized in chapter 1 by four specific technological premises and two sweeping conclusions about the future of warfare. The technological premises are that, first, improvements in computers and electronics will make possible major advances in weapons and warfare—most notably in areas such as information processing and information networks, but also in communications, robotics, advanced munitions, and other technologies; second, sensors will become radically more capable, in effect making the battlefield "transparent"; third, land vehicles, ships, rockets, and aircraft will become drastically lighter, more fuel efficient, faster, and more stealthy, making combat forces far more rapidly deployable and lethal; and fourth, new types of weaponry—such as space weapons, directed energy beams, and advanced biological agents—will be developed and deployed.

The broad conclusions that follow are that, first, if properly exploited and integrated into military organizations, tactics, and concepts of operations, these technical trends can soon add up to a revolution in military affairs that will constitute the greatest advances in warfare since the advent of blitzkrieg and aircraft carriers in the 1930s and nuclear weapons in the 1940s. This purported RMA offers great promise for the United States—affording the possibility, within ten to twenty years, of capabilities such as dominant battlespace knowledge, full dimensional protection, dominant maneuver, and precision strike from long distance.

And second, even relatively weak U.S. adversaries will greatly improve their capabilities by acquiring and learning to make good use of advanced precision missiles, satellites, antisatellite weapons, advanced mines, weapons of mass destruction, and computer viruses—and thus be able to challenge U.S. operations much more than did Iraq in Desert Storm or Serbia in Operation Allied Force. As a result, the United States will need to seek radically new military concepts to overcome these challenges to its military supremacy and, indeed, its basic security.

I suggest in chapter 1, and attempt to demonstrate in chapters 3 and 4, that the limits of physics and likely trends in engineering mean that the second and third technological premises of the RMA hypothesis are essentially incorrect. That is not to say that progress in sensor technology, as well as in major military vehicles and weapons, will be minimal, but rather that it will probably not accelerate relative to the impressive—yet nonrevolutionary—pace of the recent past. Ongoing challenges from the weather, the inherent difficulties of detecting weapons inside vehicles or buildings, and enemy countermeasures will limit the potential of future sensor technology. Slow progress in the energy content and power of explosives, the efficiency of the internal combustion engine and any possible replacement for it, the aerodynamic and hydrodynamic properties of ships and planes, and other practical realities will constrain the future performance of military vehicles and projectiles, making progress incremental in most cases.

These findings do not entirely discredit the RMA hypothesis, however. Some RMA proponents would accept them, pointing out that the first technological premise—and its implications for an information-dominant military tied together by awesome real-time computing power and data dissemination—is at the core of their claim that a contemporary revolution in military affairs is within reach.

Nonetheless, it is important to have systematically surveyed the potential of a wide range of defense technologies. Many RMA believers have much more expansive expectations about what military technology will be able to offer in the years ahead. To a large extent, moreover, they have become the prevailing RMA school of thought. This is evidenced by the heady aspirations of the Pentagon's *Joint Vision 2010* for achieving not just information superiority, but also dominant maneuver, precision engagement, full-dimensional protection, and focused logistics within a decade.

Such grandiose aspirations seem out of reach for the foreseeable future—probably well beyond 2020. And the fact that they are unrealistic has important implications for defense budgeting and resource allocation, as discussed more fully in chapter 7. Still, even if some of the specific technological premises of the RMA movement are exaggerated or wrong, and even if some of the more futuristic concepts of warfare in 2010 and 2020 do not come to pass, the nature of combat may change radically by then because of revolutionary advances in a couple of key areas of technology combined with much more modest progress in most other areas. That is, at least, a theoretical possibility.

This chapter examines that possibility. It addresses the contention of RMA proponents that the systemic, synergistic effects of combining new technologies, largely through the widespread use of modern electronics and computers, will produce staggering breakthroughs in warfighting effectiveness and, indeed, in the basic nature of war itself.

A final verdict on the RMA hypothesis is beyond the reach of military analysis—or at least beyond this author's reach. As noted earlier, there are no warfighting models or simulation techniques that do a particularly good job of capturing the capabilities and limits of existing technologies, to say nothing of technologies still on the drawing board or in the early stages of research and development (R&D).

Rather than try for an analytical knockout punch, therefore, this chapter offers various observations and predictions about a wide range of warfighting scenarios and issues. Infantry combat, naval battles, air-ground warfare, and peace operations are considered individually. In each case, some of the promises of technology between 2000 and 2020 are acknowledged, but numerous potential pitfalls and caveats are also underscored. Material from chapters 3 and 4, as well as tactical analyses that assess how various trends in individual technologies will affect warfare as a whole, are drawn upon. Several recent conflicts are examined briefly as well, with an eye toward understanding how the dimensions of warfare they highlight might change in the next two decades. They are not intended as comprehensive case studies, but simply as illustrations.

The chapter's conclusions add up to a mixed—but generally skeptical—verdict on the contemporary RMA hypothesis. It is true that battlefield information will be processed, shared, and acted upon much more quickly in the future than in the recent past. In some situations, that will produce decidely different battlefield dynamics than are expected today. However, a great deal of critical information will remain hard to acquire in the first place—in which

Box 5-1. *Summary of Key Technological Assumptions*

Before considering various areas of potential future warfare, it is helpful to summarize the key technological assumptions upon which my analysis of them rests. Most of these points have been discussed in chapters 3 and 4, but here they are organized into two main groups: projections about what technologies probably are attainable between now and 2020, on the one hand, and predictions about which will remain out of reach, on the other.

Likely Technological Accomplishments between 2000 and 2020

The following new capabilities appear likely to find their way into the U.S. military by the year 2020. Many will also be acquired by other modern industrial societies, and at least a few will be within reach of lesser-developed militaries as well.

—Computers will continue to become much faster, cheaper, lighter, and more widely used on the battlefield. Their speeds and memory capacities will increase at least tenfold, and probably much more, between now and 2020.

—Rapid dissemination of targeting data between various platforms will allow real-time network warfare; that is, weapons platforms will no longer need to acquire a target themselves before shooting. Enemy countermeasures will challenge such networks and require that they be hardened and made redundant, but they probably will not be able to shut them down.

—Unmanned sensors, coupled with increasingly powerful on-board computers, will be able to positively identify many types of distinctive military vehicles and other assets on the battlefield. By acquiring and processing a great deal of data themselves, they will largely circumvent the data transmission–rate limitations of the radio spectrum and will reduce the risks of catastrophic breakdown of a network system should central processing stations malfunction or be destroyed.

—Autonomous munitions and submunitions will be increasingly effective, at least against unsuspecting or unprotected vehicles and other large and distinctive military assets, even in poor weather conditions.

—Satellite imagery will become much more widely available. More countries will gain access to it and have at least some ability to target U.S. forces from long distances as a result. In the case of the United

States, satellites will become much more plentiful, shortening gaps in coverage of a given region or battlefield.

—Large vehicles under forest cover may be detectable by long-range sensors; in some cases, people and medium-size weapons may be detectable as well.

—Infrared-search and -homing devices will become more effective, although so will countermeasures used against them.

—Bistatic radars and infrared technologies will make some headway against stealth technology but will not be able to deny stealthy systems a certain battlefield advantage.

—Longer-range cruise and ballistic missiles will become more widely available. Their speed and stealthiness will also improve, though in a manner that will vary greatly from country to country based not only on trends in basic technology but also on arms and export control regimes and international politics more broadly.

—Most types of missiles will become increasingly maneuverable, continuing a remarkable trend under way now for several decades; manned aircraft will change less markedly in these performance parameters; unmanned aerial vehicles will be used for an increasingly wide array of missions, including the delivery of ordnance.

—Ships will become stealthier and more resilient to explosives. But on balance they will probably become more vulnerable, particularly near shorelines and in shallow waters, given trends in cruise missiles, advanced mines, and submarines.

Challenges Likely to Remain Elusive through 2020

Many other things, however, will not become possible—at least not within the first couple decades of the 21st century.

—Weapons transported inside well-sealed metal boxes or vehicles will not be detectable unless the properties of the box or vehicle itself reveal them or unless the box or vehicle can be directly approached and analyzed by an advanced short-range sensor.

—Sensors based on chemical, magnetic, seismic, or acoustic detection will improve, but their ranges will remain quite limited in most cases, and they will require reasonably close proximity to their targets to be effective.

—Small arms will be very difficult to track from long distances, as will soldiers in urban environments and, to a large extent, even those in forests.

—Nuclear and biological materials and weapons will generally not be detectable unless one has immediate access, or at least nearby proximity, to them or unless they have been brought into some type of control regime through tagging or other means.

—If large amounts of resources are not expended to harden systems against radiation, radio transmissions on the battlefield—to say nothing of the ability of computer-dependent aircraft to stay airborne and unprotected ground stations to keep functioning—may be impeded by radio-frequency weapons and high-altitude electromagnetic pulse from nuclear weapons.

—The oceans will not become "transparent" to any sensor, be it based on electromagnetic, acoustic, or another physical phenomenonology. Indeed, sonar will probably not change greatly, though enhanced signal processing and greater use of unmanned underwater vehicles may improve performance at least somewhat. At most, smaller bodies of water or localized parts of oceans might someday be transparent, in the sense of being constantly monitored through a dense sonobuoy network transmitting to satellites.

—Neither explosives nor propellants will become fundamentally more powerful per unit of weight.

—Neither jet engines nor standard internal-combustion engines will become radically more fuel efficient, lighter, or more powerful. Nor will they be superceded, for most purposes at least, by other types of engines.

—Aircraft, ground vehicles, and ships will not become radically faster or lighter.

—Sensors and munitions will not improve enough to hold deep underground targets at risk—the side hiding from the sensors and munitions will usually be able to dig deep enough to stay out of harm's way.

—Although some progress will be made against snipers and other irregular threats, enemy soldiers operating in small groups and using small weapons and sound tactics will often be able to get in the first shots against any intervening forces, be it with rifles, grenades, antitank weapons, or man-portable antiaircraft weapons.

—Hypervelocity and space-based projectiles may become feasible, but they will be limited in number and effectiveness due to modest payloads and high costs.

—Nonlethal weapons will reduce casualties on the battlefield, but will generally be unable to neutralize enemy forces that are beyond the range of lethal weapons.

case there will be little that even an extremely fast real-time information grid can do to compensate. In addition, long-range strike systems will have important limitations, except perhaps against the most exposed targets.

Infantry Combat

What will technological and tactical innovation do to warfare in urban, mountain, forest, and other complex settings where airpower and large mechanized vehicles have traditionally been of limited usefulness? This type of warfare tends to involve smaller units, with battles taking place on smaller land areas than in major armored war.

This category of operations is important to consider—even for those generally opposed to using the U.S. military to intervene in civil wars or stop genocides—as it could also be relevant to strikes against terrorist groups, searches for missing fissile materials or other critical substances, and efforts to overthrow an extremely dangerous regime within its capital city.

Better technology, and associated breakthroughs in tactics and operations, will certainly help U.S. forces in infantry combat settings. Indeed, they have been for decades. Night-vision technology, helicopter mobility, advanced armored vehicles, and body armor are among the key capabilities that have led to recent American successes in infantry battles. The 1989 invasion of Panama was a remarkable success, quickly convincing the Panamanian defense forces not to mount a protracted resistance. In that conflict, U.S. losses were roughly one-third as great as those of Panamanian soldiers when fighting did occur. In Somalia in 1993, despite the ultimate failure to achieve the operation's more ambitious goals, individual battles went very much to the advantage of American troops—on the order of 10 to 1 or more in casualty ratios, for example.[1] When fighting from tactical defensive positions, U.S. infantry forces have sometimes done even better. For example, in Desert Storm a special forces team deployed into Iraq to conduct reconnaissance for the ground campaign was discovered, but fortunately had time to prepare for battle before Iraqi troops could arrive. In the ensuing firefight, roughly 150 Iraqis were killed and no Americans were lost.[2]

1. "U.S. Military Operations in Somalia," Hearings before the Senate Committee on Armed Services, 103 Cong. 2 sess. (Government Printing Office, 1994), pp. 8–11, 29–35, 45–53, 62–63; "1989 Events in Panama," Joint Hearings before the Senate Armed Services Committee and Select Committee on Intelligence, 101 Cong. 1 sess., October 6 and 17, December 22, 1989 (Government Printing Office, 1990), p. 133.

2. Susan L. Marquis, *Unconventional Warfare: Rebuilding U.S. Special Operations Forces* (Brookings, 1997), pp. 235–36.

These combat results do not suggest that technology is a panacea for infantry warfare, however. U.S. forces suffered considerable losses in Somalia and Panama, even if they were much less than those of their adversaries. Relatively unsophisticated, but proficient and disciplined, adversaries could probably do even better by fully exploiting tactics such as ambushing. For example, lightly armed units playing the role of enemy or opposition force at the Joint Readiness Training Center at Fort Polk, Louisiana, routinely do very well against the better-armed army infantry units practicing against them. The home-base opposition force in these exercises routinely inflicts more casualties (as determined by laser-activated equipment that simulates the effects of actual weapons) than it suffers itself.[3]

Infantry Combat in 2020

What of these various realities might change substantially over the next one to two decades? Some major improvements will be possible for the United States and other advanced militaries; however, several major problems for a high technology force will endure.

Battlefield transport is unlikely to change rapidly in the next twenty years. Fixed-wing transport aircraft will change relatively little. Helicopters will become somewhat faster and stealthier and include more tilt-rotor technology. But the stealth revolution is unlikely to make the infantry battlefield safe for planes or helicopters—particularly transport aircraft—the way it has enhanced the prospects of high-flying attack jets. Armor on platforms such as infantry fighting vehicles will become somewhat tougher and lighter. Vehicles are likely to keep becoming more safe, and protective measures are likely to improve against land mines—at least against the simple types of mines and other threats available today. New materials and the better use of existing ones, like Kevlar, will make vehicles more resistant to any land mines they might encounter and detonate; much better types of mine-resistant vehicles, such as those designed by South Africa and Australia, will become more widespread as well.

Enemy forces will possess some advantages, however, against these types of vehicles. Many types of weapons relying on passive infrared sensors—including everything from small antitank devices to surface-to-air missiles—will remain very hard to find and attack before they can get off the first shot themselves.[4] This is troublesome—and perhaps especially so for the threat

3. Michael O'Hanlon, *Saving Lives with Force: Military Criteria for Humanitarian Intervention* (Brookings, 1997), pp. 71–72.

4. Nick Cook, "Serb Air War Changes Gear," *Jane's Defence Weekly*, April 7, 1999, pp. 1–4.

posed by surface-to-air missiles—since infantry warfare will continue to require countries like the United States to fly large numbers of people to and around the battlefield, placing many of its personnel at risk.

Adversaries may also have access to improved mine technology. Their mines may be networked, allowing simultaneous detonation once a convoy or column of vehicles has entered a minefield; presently, such minefields would typically harm only the lead vehicles in such formations before others realized the danger. Adversaries may also possess standoff mines, like the U.S. wide-area mine, which can sense a target tens of meters away and then attack it with direct-attack or homing munitions. Among other advantages, such mines could provide their users with minefields that are much harder to sweep or detect visually.[5]

Even when enemy weapons are detected by reconnaissance assets when fired, their ability to "shoot and scoot" may make it impossible for standoff weaponry to reach them before they can hide in buildings, crowds of people, or other complex terrain. Recent examples of this phenomenon include Iraq's SCUD missiles and the mortars of militias in Mogadishu. Suppose the air force succeeds in its goal of developing sensitive infrared detectors that could spot the firing of a mortar or artillery tube.[6] Even so, there would still be a delay of at least several seconds in getting ordnance (or a nonlethal weapon) on the target. That could be long enough for a mortar or artillery tube based on a specially designed small truck to be mechanically retracted from view and hidden below the truck's metal skin. The skin could provide the mortar protection against a radio-frequency weapon. In urban combat settings, the truck's appearance might make it indistinguishable from nearby civilian vehicles. If the enemy truck had been tracked continuously by an aircraft or unmanned aerial vehicle (UAV) after firing its shots, it could be attacked with conventional explosive or possibly a nonlethal agent. However, if the vehicle could shelter itself behind a building, such a continuous track might not be possible to establish and maintain. Moreover, trends in man-portable air-defense systems may make such types of airborne monitoring assets increasingly vulnerable to preemptive fire, day or night.

Countersniper technologies will make it even more likely that enemy riflemen, as well as immediate accomplices, will be quickly located and neu-

5. Robert M. Allen, "Future Land Mines," in Defense Science Board 1996 Summer Study Task Force, *Tactics and Technology for 21st Century Military Superiority*, vol. 3, *Technology White Papers* (Department of Defense, 1996), pp. V-111 through V-121.

6. Robert Wall, "USAF Tackles Urban Combat," *Aviation Week and Space Technology*, March 22, 1999, p. 83.

tralized. However, those enemy riflemen will still generally be able to get the first shot in a given setting, since planned countersniper technologies will not work until an enemy has already fired or at least aimed its weapon. Good riflemen will also plan their shot, camouflage, and escape route carefully, minimizing their vulnerability to reprisal.[7]

Small, guided, precise mortar rounds may become available to U.S. forces, making an individual sniper and anyone accompanying him less able to take cover. But they will generally not make it possible to take out more than one target at a time, since independent information about each enemy weapon will still be needed. Snipers can also be expected to be more selective about where they fire from, and how long they aim, as countersniper technologies improve. For example, they may use the cover provided by schools and hospitals more frequently or prepare sites inside other buildings with sandbags and other means of reducing their own vulnerability to countersniper fire.[8]

Nonlethal weapons will help in some situations—notably in infantry operations intended to bolster a peace accord. To be effective, however, nonlethal weapons will generally require a good line of sight to a target and reasonably close proximity—meaning that they will generally work only in situations where lethal weapons would also work and, in that regard, not constitute a major breakthrough for high-intensity combat operations. They may be important for saving lives. But they will be unlikely to affect the basic course of battle.

Improvements in computing speed may allow more signals from enemy radios to be deciphered, even if they are buried in noise or encrypted. However, the basic sensors picking up these signals are unlikely to improve dramatically. In addition, encryption technology will improve, and adversaries may eventually use broadband radio that is harder to locate. Perhaps even more to the point, human networks and internal lines of communication using fiber-optic cable or other fairly secure media will allow many adversaries to transmit a good deal of information back and forth without using radios.

One implication of these observations is that strikes such as the 1998 U.S. attack against terrorist Osama bin Laden—which failed to kill bin Laden or deter him from attempting subsequent attacks—will probably not

---

7. John Miller and John S. Eicke, "Counter-Sniper Technology," in Alex Gliksman, ed., *Meeting the Challenge of International Peace Operations: Assessing the Contribution of Technology* (Livermore, Calif.: Center for Global Security Research, Lawrence Livermore National Laboratory, 1998), p. 144.

8. George I. Seffers, "Next U.S. Army Rifle Will Attack around Obstacles," *Defense News*, January 12, 1998, p. 13.

get much easier with time.[9] Targeting information will remain hard—and perhaps become harder—to obtain as individuals such as bin Laden come to recognize the capabilities of U.S. signals intelligence. Human intelligence networks may well continue to prove the most practical, yet also the most nontechnological, way of finding such people.

Communications for U.S. forces will improve. One of the most useful innovations will be greater use of UAVs to provide backup to satellites, ensuring greater resilience of the communications network against the effects of antisatellite weapons and other disruptions.[10] In terms of new capabilities, the chief breakthrough in communications will probably be that information will be shared much more quickly. The concept of a tactical internet will become more and more of a reality for all types of ground forces, including infantry soldiers, who will be able to obtain very good real-time information on each other's whereabouts. In addition, whatever information they can gather on the enemy will be available to a wide range of troops quickly. Robotics and small UAVs, acoustic sensors, voice-recognition software, and other improvements in sensor technology will greatly aid in their efforts to gain such information.[11]

However, most of these sensors require some prior general information on where enemy forces might be operating; their expected ranges are on the order of meters or tens of meters in most cases, not kilometers. Even if they can penetrate one wall, as in the case of advanced acoustic sensors, they may not be able to penetrate five or ten. Adversaries can also be expected to take note of the increased vulnerability of any single site and disperse their assets more widely; for example, they may minimize the number of times key leaders assemble in a place where they would be highly vulnerable if located.

To be sure, certain situations will allow robotics, other sensor improvements, enhanced computing power, and other developments to make major changes in how infantry battles can be fought. For example, troops in a deciduous forest during wintertime may be targeted by munitions with advanced infrared sensors together with advanced software that allows computers to search for signatures that move from frame to frame. Some could be hit by shrapnel even if they were partially protected by trees. Alterna-

9. On the bin Laden network's subsequent attempts at further strikes, see Vernon Loeb, "Has the U.S. Blunted bin Laden?" *Washington Post*, February 17, 1999, p. A4; Ivan Eland, "Does U.S. Intervention Overseas Breed Terrorism?" Foreign Policy Briefing 50 (Washington, D.C.: Cato Institute, December 17, 1998).

10. *Defense 97: Pathways to the Future* (Department of Defense, 1997), p. 31.

11. See Michael G. Vickers and Robert C. Martinage, *The Military Revolution and Intrastate Conflict* (Washington, D.C.: Center for Strategic and Budgetary Assessments, 1997), pp. 19–20.

tively, enemy soldiers known to be within an otherwise deserted city block or building will have a wide array of technologies aimed at locating them. But a high-technology military will still need to know the general vicinity in which to search for enemy forces and to have weaponry in position to attack them once they are found.

Foliage-, soil-, and single-wall-penetrating radars may become workable in the next decade or two. However, considerable technical obstacles still prevent them from being practical, and even if those challenges are surmounted, the new sensors will have limits. They will not be able to see through multiple walls or penetrate the trunks of trees, meaning that objects hidden deep within buildings or immersed in dense forest may still evade detection, or be difficult to track and attack even if they are momentarily spotted. For these reasons, U.S. soldiers operating in heavy forest or urban environments will be able to call in outside fire against only some targets and will probably have to fend for themselves in many situations. This, in turn, means that they will probably still have to deploy in enough strength to take care of each other, and not simply to play the role of scouts or target designators for standoff attack.

In general, clever and entrepreneurial enemies—even if relatively poor in economic terms—will remain able to thwart many advanced technologies of an adversary, particularly when fighting on their own terrain. Many of the tactics used by North Vietnamese and Viet Cong forces will remain useful in future warfare: using tunnels and bunkers, "hugging" enemy forces to provide a form of protection from long-range wide-area firepower, storing materials in small amounts throughout a battlefield, employing large labor forces to transport supplies in modest amounts, and making wide use of pontoons and underwater bridges.[12] Many of these tactics will remain useful even against a foe with foliage- and soil-penetrating radar, other advanced sensors, and brilliant anti-armor weapons.

These observations and predictions can be summarized another way: in infantry combat, ambush will remain possible even for a technologically outclassed force. In forests and even more so in cities, where cover is plentiful, the world's less advanced militaries will continue to have access to weaponry with greater ranges than a top-notch military's sensors. If they use good tactics, they will often be able to get in the first shot. It is for such reasons that recent Marine Corps experiments testing new technologies in urban settings found that the high-tech forces still would suffer extremely

12. Jeffrey A. Isaacson, Christopher Layne, and John Arquilla, "Predicting Military Innovation," briefing, RAND Corporation, Santa Monica, Calif., 1999, pp. 36–37.

high casualties.[13] This stands in contrast to the situation when, for example, the Spanish explorer Francisco Pizarro conquered the Incan emperor Atahuallpa in the sixteenth century; in such engagements, no more than a few hundred Europeans defeated many thousands of native Americans, whose weapons were almost entirely ineffective against horseback-mounted, steel-wielding opponents.[14] It also stands in contrast to Operations Desert Storm and Allied Force, during which coalition aircraft could often attack enemy armor while staying out of harm's way themselves.

These realities about infantry combat, which are likely to endure well into the future, have convinced the army's chief long-term planner to propose an indirect approach to future urban operations. Following the spirit of British strategist B. H. Liddell Hart's writings, U.S. forces would avoid combat in cities as much as possible, instead quarantining it or otherwise putting pressure on enemy forces asymmetrically.[15] Unless a city's entire population is to be treated as the enemy, however, or unless militias are too weak to be able to keep populations hostage within city boundaries, quarantining the city may not be practical.

Case Study: Mogadishu, 1993

The October 1993 firefight between U.S. forces and the partisans of Somali warlord Mohammed Farah Aideed is an important case study of infantry battle. In that conflict in which eighteen Americans and several hundred Somalis died, the United States did not lack good information about the location of enemy centers of gravity. It had correctly identified the site of a meeting of key Aideed lieutenants. In other words, it did not lack for good and timely intelligence. As retired Marine Corps lieutenant general Paul K. Van Riper put it, "We had information dominance in Somalia."[16] The United States got into trouble, rather, when helicopters transporting assault forces to exactly the right place for their mission were unexpectedly shot down.

Given trends in antiair missiles and helicopters, this same problem could easily recur in 2020. Having better intelligence on where enemy forces were

13. Thomas E. Ricks, "Urban Warfare: Where Innovation Hasn't Helped," *Wall Street Journal*, October 12, 1999, p. 10.

14. Jared Diamond, *Guns, Germs, and Steel: The Fates of Human Societies* (W. W. Norton, 1997), pp. 67–75.

15. Robert H. Scales Jr., "The Indirect Approach: How U.S. Military Forces Can Avoid the Pitfalls of Future Urban Warfare," *Armed Forces Journal International*, vol. 31 (October 1998), pp. 68–74.

16. See James Kitfield, "The Air Force Wants to Spread Its Wings," *National Journal*, November 8, 1997, p. 2265.

located would, as noted, have made little difference in Mogadishu. Although U.S. forces were surprised by the number of hostile Somalis who converged on them, guns in hand, once the fighting began, no RMA technology in development would have made it possible to predict such a semi-spontaneous gathering or to locate all the arms stored in individual households that made the crowd so dangerous.

Nonlethal weaponry would have helped some, but the notion of having an areawide acoustic weapon or some other means of immobilizing most people when U.S. forces went in to make their arrests presupposes technological advances that are by no means certainties before 2020. Nor would stun guns and other such nonlethal direct-fire weapons have saved many American lives. Although Somalis were inclined to use civilians as human shields when attacking, U.S. forces generally fired back anyway under those hazardous circumstances.[17] Nonlethal weapons might have saved some Somali lives, in situations where noncombatants could be clearly identified, yet they would not have altered the battle's basic nature.

Better knowledge of Somali city streets made possible by the continued spread of global positioning system (GPS) receivers and computerized databases on U.S. military vehicles would have helped, as would have more resilient and agile armored vehicles. Still, many Americans died not on Mogadishu's streets but in firefights at the original sites where the helicopters went down.[18] On balance, if one imagines that battle being fought in 2020 instead of 1993, it is doubtful that trends in technology and associated warfighting tactics in the intervening years would radically alter its dynamics or outcome.[19]

Air-Ground Warfare and Heavy Combat

Warfare involving heavy weaponry and aircraft is likely to change more than infantry battle in coming years. But even here major questions and limitations exist concerning the potential of innovations made possible by new technology.

It is worth noting that the aforementioned realities of infantry combat are relevant to heavier warfare as well. Many high-intensity wars include,

17. See Mark Bowden, *Black Hawk Down* (Atlantic Monthly Press, 1999), pp. 78, 99, 106.
18. Major General William F. Garrison and Major General Thomas Montgomery, statements before the Senate Committee on Armed Services, 103 Cong. 2 sess., May 12, 1994 (Government Printing Office, 1994), pp. 8, 27, 35–36, 53–55.
19. See Bowden, *Black Hawk Down*, pp. 1–197; Garrison and Montgomery, statements before the Senate Committee on Armed Services.

or conclude with, phases of urban and infantry fighting, particularly if an objective in the war is to overthrow a government and occupy a country. So the first and broadest point to make is that, depending on the political and strategic objectives, there may not be many future wars made up purely of high-intensity armored battles.

In brief, the state of heavy combat at the turn of the century might be summarized as follows. Within ranges of a few kilometers, lethalities are considerable. Tanks can see and accurately target each other from such distances; many antitank weapons work at that range, too. Aircraft can distinguish armored vehicles from trucks and other platforms, and use laser-guided bombs and Maverick missiles against them, at several kilometers' range as well. These munitions can be quite effective. For example, NATO's 1999 aerial attacks against Serbian forces in Kosovo reportedly destroyed about 90 tanks, 150 armored combat vehicles, and almost 400 artillery pieces and mortars, using a total of several thousand munitions; even if the true attrition totals are only half as great as advertised, the overall lethality is very impressive by the standards of pre–Desert Storm warfare.[20] As for aerial combat, aircraft are now capable of recognizing and destroying each other from twenty or more kilometers' distance.

Reconnaissance platforms can see armor—day or night—at many tens of kilometers. They can do so via overhead optical satellites, in which case vehicles can often be recognized, or from airborne moving-target identification systems like the joint surveillance and target attack radar system (JSTARS) aircraft, in which case they are harder to recognize but can be seen even in bad weather.

There are important limits on sensors, too, however. Tactical radars often cannot tell tanks from trucks beyond a few kilometers' distance. The ranges of precision weapons are not that great. Moreover, those that are truly precise are severely hampered by poor atmospheric conditions; all-weather GPS- and radar-guided weapons are "near precise," at best, with typical miss distances of six to twelve meters rather than one or two.

Although technologies like infrared-vision devices, laser range finders, and stabilized guns allow Western technology to be more effective than its main rivals, even less advanced tanks and antitank weapons are capable of destroying U.S. and European armor if they get a good shot.

---

20. See General Wesley Clark, "Kosovo Strike Assessment," NATO Headquarters, Brussels, Belgium, September 16, 1999.

Air-Ground Warfare and Heavy Combat in 2020

Trends now under way will "stretch" the battlefield and also permit greater weapons effectiveness in poor weather conditions. The stretching of the battlefield is not surprising in one sense; it simply perpetuates a trend that has been under way for centuries, as can be seen from figure 5-1. In antiquity, roughly 500 soldiers filled up the battlefield space equivalent in size to one football field; in the American Civil War twenty took up that space; in World War I, just two did. In World War II, a single soldier occupied roughly five football fields, and, according to 1980s NATO doctrine, one soldier would have had ten football fields to himself.[21]

These trends will continue. Precision weapons will continue to increase in range. In one sense, their range is already substantial, as demonstrated by cruise missiles capable of flying hundreds of kilometers. But these long-range weapons are not capable of truly precision strike; their accuracies are generally measured in the low tens of meters, and they do not have the terminal homing capabilities needed to destroy moving targets. In the future, they will be able to drop advanced submunitions using infrared sensors that operate below many types of cloud cover and, in some cases, acoustic sensors or semi-active radar as well. Over the longer term, autonomous radars may be placed on munitions and submunitions, making them effective in virtually all weather conditions. More weapons may be built that use GPS sensors or inertial guidance to fly into the general vicinity of a target and below most types of cloud cover, at which point a television camera on the weapon can begin sending images back to the launching plane's cockpit for terminal guidance (as with today's AGM-130).

In addition, improved sensors will provide targets for these weapons from greater ranges. A host of technologies, like UAVs and tactical data links, will allow one platform to see a target and another to shoot at it. This situation has already been realized insofar as an aircraft like JSTARS might pass along target coordinates to a fighter or attack helicopter. The shooter platform usually still needs to get its own track on the target before shooting, however, or have another aircraft nearby or a person on the ground laser-designate for

---

21. To put it differently, in antiquity 100,000 men filled up a single square kilometer, and by World War II they occupied nearly 3,000 square kilometers. See Trevor N. Dupuy, *Attrition: Forecasting Battle Casualties and Equipment Losses in Modern War* (Fairfax, Va.: Hero Books, 1990), p. 28; Trevor N. Dupuy, *Understanding War: History and Theory of Combat* (Paragon, 1987), pp. 172–73.

Figure 5-1. *Historical Army Dispersion Patterns[a]*

Typical spacing between soldiers (meters)

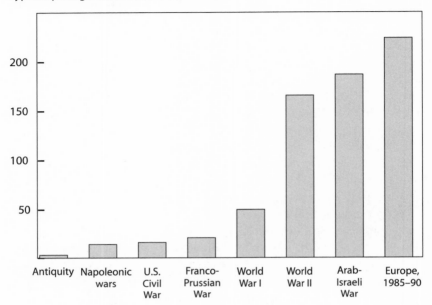

Source: Trevor N. Dupuy, *Attrition: Forecasting Battle Casualties and Equipment Losses in Modern War* (Fairfax, Va.: Hero Books, 1990), p. 28.

a. Army or corps of 100,000 troops.

it. This is inconvenient in many cases and dangerous in some. In the future, data links will certainly be much faster and more capable.[22]

Guidance technologies and submunitions now becoming available will also translate artillery from an indirect-fire, imprecise-area weapon to a weapon capable of routinely destroying armor. Recent U.S. Army simulations suggest that roughly half of all tank kills in future armored combat may come from such rounds. Two important such systems are the brilliant antitank munition (BAT) and the sense and destroy armor munition (SADARM), which are focused mostly on softer armored vehicles such as self-propelled artillery. Both use infrared sensors, while the former also employs acoustic sensors and the latter, millimeter-wave radar homing. They may achieve this type of standoff anti-armor capability in the next few years, assuming they work as planned (initial results appear promising) and do not prove unduly vulnerable to straightforward countermeasures.[23]

22. Cook, "Serb Air War Changes Gear."

23. Institute of Land Warfare, "The Division Advanced Warfighting Experiment: Fire

Angular stealth coatings may change the external appearances of some combat vehicles. In some situations, as noted earlier, countries may try to make their heavy weapons look like civilian vehicles from the outside. Tanks may be somewhat smaller. They may have electromagnetic guns but are perhaps more likely to have more efficient explosives and a greater variety of rounds to fire.

A number of things are quite unlikely to change. Armor itself will involve a somewhat different mix of materials, but it will continue to rely at least in part on sheer density and weight for protection, even if the roles of stealth and countermeasures increase. Various armored vehicles will remain prevalent on the battlefield to provide speed of maneuver, protection against indirect and direct fire as well as mines, and space to hold the computers and communications devices and munitions that soldiers will need in order to be lethal against most large and distant objects. All of these vehicles are likely to continue to rely on internal-combustion engines powered by petroleum products, so fuel and logistics requirements for significant combat forces will remain quite high.

Even assuming rapid technological progress in many areas of heavy combined arms warfare, it will not always be straightforward to radically revise warfighting tactics and doctrine. Consider, for example, the notion that high-technology scout teams may be able to substitute for large formations of ground forces in some situations, using a wide array of technologies to create real-time databases about enemy forces that can then be used by stand-off weaponry operating from rear areas. Such scout teams may prove quite capable. However, this scout-team concept may not work in all situations. In warfighting experiments involving such small teams so far, opposition forces have elected not to pursue the elusive teams unless they experienced acute attacks that were probably being directed by them. Instead, they have interdicted battlefield resupply of the teams—in effect, starving them out over time.[24] Another option would be for large forces to use their own small teams of foot soldiers to protect their flanks against enemy scout teams, paying a bit of a premium for protection but making the scout teams' jobs much more hazardous and difficult. Finally, the scout teams might also prove vulnerable if they could be identified by their radar or radio emissions or some

---

Support Implications (Crusader, MLRS and Comanche)," Defense Report 98-3 (Arlington, Va.: Association of the U.S. Army, April 1998).

24. Joe Polito and Dan Rondeau, "Systems Concepts for Relatively Small, Rapidly Deployable Forces," in Defense Science Board 1996, *Tactics and Technology for 21st Century Military Superiority,* vol. 3, pp. V-1 through V-5; Elaine M. Grossman, "Experimental Squads, Supported by Air, Provide Effective 'Force Multiplier,'" *Inside the Pentagon,* April 24, 1997, pp. 5–6.

other such signature. Armored vehicles might help protect them under such circumstances. But in that case, logistics demands on the scout teams would go up considerably, and their stealthiness would be compromised.

On the future battlefield, many fixed or semi-stationary assets like airfields, depots, barracks, ports, and even ships will probably be much more vulnerable than they are today to standoff cruise missile attack, as well as to any precision munitions with terminal homing based on light or radar that enemy special forces may operate. Vulnerabilities may be mitigated by greater dispersal of forces and hardening of key military assets, together with improved missile defenses and countermeasures against attacking missiles. Greater use of vertical and short takeoff and landing airplanes that do not require long runways may also help.[25] But these may prove only partial solutions, and such trends may complicate a high-technology force's operations and limit its dominance.

Long-range electronics and communications on which these advanced air-ground combat operations would depend may not prove robust. Anti-satellite weapons, electromagnetic pulse (EMP), and radio-frequency weapons as well as computer viruses and hacking may challenge them. By spending money hardening electronics and building redundant systems, a country like the United States can probably overcome most such threats to its advanced military—but doing so will not come cheaply and may be less than 100 percent reliable.

Finally, enemy countermeasures may prove able to defeat homing submunitions. It is likely that improved precision weaponry will still make precision engagement an increasing reality, but net progress may not be as rapid as many assume. Even if decoys, flares, noisemakers, and other such devices cannot effectively mimic the effects of real targets and lead munitions astray,

25. Airfield vulnerability has sometimes been exaggerated in the past, as Joshua Epstein demonstrated in an important 1984 book on the Soviet threat to NATO airfields. Among the reasons why airfields are difficult to incapacitate is the fact that they are large; standard doctrine would require more than 200 hits with conventional ordnance to disable the airfield by essentially cutting it in two and attacking a number of the taxiways, hangars, exposed aircraft, and other assets. However, some of these attacks are essentially redundant if the goal is to shut down an airfield; in that case, only about 100 (accurate) bombs would be needed. In fact, a smaller number of munitions may do much to reduce the usage of an airfield—even if complete suppression is deemed to require such a large number, an airstrip of only fifty meters' width could be made unsafe with several craters along a single line perpendicular to the long axis of a runway—meaning that even ten properly spaced munitions might do considerable damage. See Joshua M. Epstein, *Measuring Military Power: The Soviet Air Threat to Europe* (Princeton University Press, 1984), pp. 136, 199–201; Thomas A. Keaney and Eliot A. Cohen, *Gulf War Air Power Survey Summary Report* (Government Printing Office, 1993), pp. 42, 63, 185.

it will often be possible for an adversary to build many armored vehicles to resemble civilian vehicles, making it quite hard for the United States to know what to attack. This approach may not work in a place like the desert of southern Iraq and Kuwait, where civilian traffic as well as urban and forest cover are sparse, but it could work well in more complex terrain and in places where military vehicles would be interspersed with large amounts of civilian traffic.

Some might describe these likely improvements in enemy capabilities and tactics as part of the so-called revolution in military affairs. However, it is equally valid to describe them as impediments to a true revolution.

Not all obstacles to an RMA in high-intensity warfare are military or technological in nature. Some are political as well. Countries with RMA technologies may be strategically fooled or paralyzed, thereby failing to bring their advanced systems to bear against an enemy at a time when the enemy is most vulnerable. To fool or confuse U.S. intelligence and policymakers, an enemy might conduct a major military buildup against an American ally very gradually. It might disguise the buildup as an exercise. It may otherwise try to conceal its true intent by interspersing military vehicles within civilian convoys or traffic. In these ways, when the enemy finally attacks, its acute military exposure may be limited to a short period. Historically, these types of tactics tend to work (most recently when Saddam Hussein invaded Kuwait in 1990).[26] Nothing about the purported RMA will make these types of political dilemmas easier—and faster—to resolve. Even the most advanced weapons will need a certain number of hours, and more commonly a few days, to reach a distant overseas theater and undertake a large-scale counterattack. That will sometimes be too slow.

Case Study: Desert Storm, 1991

Consider how all of these considerations and trends, taken together, might have affected Operation Desert Storm, and particularly the matter of U.S. casualties in that war. Nearly 400 Americans lost their lives over the course of the buildup and war: almost 100 were lost during deployment and training of forces before hostilities began, about 150 were killed in action (thirty-five by "friendly fire"), and another 150 died from encountering unexploded ordnance and mines or suffering accidents during the war.[27]

26. See Richard K. Betts, *Surprise Attack* (Brookings, 1982); Lawrence Freedman, *The Revolution in Strategic Affairs*, Adelphi Paper 318, International Institute for Strategic Studies (Oxford University Press, 1998), pp. 41–45.

27. See Department of Defense, *Defense Almanac 96* (Alexandria, Va.: American Forces

Regarding those thirty-five killed accidentally by their own forces, future military technology should be able to prevent most such losses. Assuming information grids are not brought down—something an adversary of the caliber of Iraq would probably have difficulty accomplishing—all U.S. forces should know where their compatriots-in-arms are in real time. The United States should have reliable identification-friend-or-foe (IFF) devices on individual pieces of equipment, as well as information networks that display data on the whereabouts of friendly vehicles and even individual friendly troops.

Many in the group of 150 casualties who died from accidents, and some in the group of 100 who died before hostilities erupted on January 17, might also have been saved by advanced technologies (and in some cases just by better use of existing technology). Safer and more mine-resilient vehicles could reduce the dangers from mines; more advanced equipment diagnostics could continue the trend in improved safety that has reduced U.S. military death rates due to accident by more than 50 percent in the last two decades. These types of changes, however, will be gradual, as they have been in the past, perhaps resulting in improvements of 10 to 50 percent by 2020. Taking these projected improvements together, the United States might save fifty to 150 lives out of the 400 lost in Desert Storm.

Other trends are not as promising. The spread of UAVs and reconnaissance aircraft, as well as real-time date processing and dissemination, will help U.S. forces. But trends in infrared search technologies and other sensors will help even lower-technology forces ambush an attacker using anti-tank and antiair weapons. Heavier armor and new protective materials promise no magic solutions to these types of vulnerabilities. On balance, technology might reduce U.S. losses in such a future war in some ways but increase them in others. In fact, a foe armed with modestly better equipment than Iraq possessed in 1991 could easily cause the United States more casualties in a 2020 replay of a Desert Storm–like war than it suffered in the actual war. That could be the case even if the adversary did not shoot down a transport airplane, sink a troop ship, use weapons of mass destruction, or otherwise radically change the course of the war.

Consider this simple calculation that illustrates the limits of high technology, even in a war like Desert Storm. In typical ground battles in Desert

---

Information Service, 1996), p. 44; Department of Defense, *Worldwide U.S. Active Duty Military Personnel Casualties: October 1979 through June 1997* (Washington Headquarters Services, Directorate for Information Operations and Reports, 1997), pp. 1, 7, 33; Michael R. Gordon and Bernard E. Trainor, *The Generals' War: The Inside Story of the Conflict in the Gulf* (Boston: Little, Brown, 1995), p. 457.

Storm, such as the Battle of 73 Easting, Iraq suffered roughly twenty to 100 equipment or personnel losses for every U.S. loss. But, as Stephen Biddle has estimated, Iraq would have done much better—perhaps even fighting the United States to a draw there (and thus in similar battles, presumably)—by properly digging its tanks into prepared positions and making better use of advance guards. Biddle does not reveal the intricacies of his model, but the following simple dynamic is roughly consistent with his results: If Iraqi units were properly dug in and forewarned, they would generally get in the first shot at approaching U.S. forces. Assuming those first shots were lethal between roughly one-quarter and one-third of the time, Iraq might wind up causing about as many American losses as it suffered. (In situations where the Iraqis' first shots missed, the United States would generally—but not always—win the ensuing exchanges, resulting in a roughly equitable exchange ratio between the two forces overall.)[28]

Now assume a year 2020 RMA environment. What might change in the Battle of 73 Easting, assuming it could be refought between U.S. and Iraqi forces at that time? This question involves too many complexities and unpredictable features to answer rigorously; however, one simple way to think about the main dynamics might be this: assume that U.S. information-gathering networks would be so thorough and fast that enemy armor would remain as exposed as Iraq's was in 1991, even if properly dug in. That may be a generous assessment of U.S. reconnaissance and weapons capabilities, but it is not entirely implausible.

Nonetheless, it is quite unlikely that in 2020 infantry soldiers holding man-portable antitank weapons would be spotted by satellite, aircraft, or UAV. They could generally stay completely out of view in foxholes until U.S. forces approached, particularly if foliage or buildings provided cover. In many cases, they would be able to ambush U.S. forces.

It is impossible to predict the precise quality of enemy antitank weapons in 2020. For purposes of illustration, however, assume that they would each have a 10 to 20 percent chance of destroying a nearby tank, after which they themselves would be vulnerable to counterfire. (This range of possible weapons effectiveness is not implausible. After all, even simple antitank rockets can destroy today's best tank if fired from close range, and tube-launched, optically tracked, wire-guided [TOW] missiles were fairly effective against Iraqi armor in Desert Storm.)[29] Under these circumstances, the enemy force

28. Stephen Biddle, "Victory Misunderstood: What the Gulf War Tells Us about the Future of Conflict," *International Security*, vol. 21 (Fall 1996), pp. 146–47, 162, 167.

29. Biddle, "Victory Misunderstood," p. 164; Department of Defense, *Conduct of the Persian Gulf War: Final Report to Congress* (April 1992), p. T-129.

firing first would, if holding one antitank weapon for every approaching U.S. armored vehicle, destroy roughly one American combat vehicle for every three to five losses of its own antitank positions. Relative personnel losses would probably be comparable. This outcome would be lopsided and in the favor of the United States, but still nearly ten times worse than the exchange ratios achieved in Desert Storm.

Peace Operations and Humanitarian Interventions

How might trends in military technology and associated doctrinal innovation affect peace operations and forcible humanitarian interventions over the next twenty years?

These types of missions are closely linked to the category of infantry warfare, because their most challenging elements are often small-unit combat operations involving light weaponry. For example, the 1993 Mogadishu firefight considered previously occurred in the context of a humanitarian military intervention. That experience illustrated the adage that humanitarian and peace operations are bad missions for soldiers, but only soldiers can do them. In other words, peace operations and forcible humanitarian interventions may be a distraction for the armed forces of countries like the United States. But the fact that they can involve intensive fighting means that those conducting them must possess well-honed, traditional combat skills.

Peace operations and humanitarian interventions also overlap considerably with infantry missions in that they often require policing, securing key infrastructure, and providing supplies to troops as well as civilian populations. These tasks are all assigned to infantry units in certain kinds of wars, particularly those culminating in the occupation of territory. For defense planning purposes, important distinctions are made between infantry combat operations and peace operations since each places different relative strains on a military force structure. For example, it may be appropriate to place most military police units in the reserves if they are most likely to be needed in the rare war of occupation, but it may be wise to place some in the active-duty force structure if they are to be used in frequent peace operations. From the perspective of basic military competencies and technologies, however, the missions are very similar.

Has airpower also become a primary instrument of humanitarian intervention? NATO's 1999 war against Serbia in defense of Kosovar Albanians might make one think so. After all, NATO did manage to force Yugoslav president Slobodan Milosevic to capitulate without invading Kosovo with ground troops. That led to a reversal of the wide-scale ethnic cleansing at a relatively modest cost in lives to ethnic Albanians, Serbs, and its own troops. Too many lives were still lost, and NATO's initial goal of reducing the scale of atrocities against Kosovar Albanians was clearly not realized.[30] Ground units of the Kosovo Liberation Army (KLA) played a role in the military victory as well.[31] Still, that NATO's victory was less than absolute, and somewhat dependent on good luck as well as help from the KLA, does not change the facts that the outcome was reasonably good (as the outcomes of wars go), and that airpower deserves most of the credit for it.[32]

NATO's war against Serbia was rather unusual for this type of conflict, however. Most civil wars occur in less-developed countries such as in Africa, where armored forces susceptible to attack from the air are few and far between; they also commonly involve roving militias that are difficult to influence through strategic attacks against fixed infrastructure. Finally, airpower worked against Serbia only when Russia tightened the diplomatic noose around Slobodan Milosevic by essentially accepting and promoting NATO's demands and when NATO sent increasingly clear signals that it was considering a ground invasion if necessary.[33] So, on the whole, while airpower may have a certain role to play in some humanitarian interventions, infantry-style operations are generally much more central to them.

Peace Operations and Humanitarian Interventions in 2020

Technology offers only so much potential to change peace operations and humanitarian interventions. It cannot resolve the political challenges that outside countries always face in conducting these types of missions, such as whether to take sides in a conflict, simply create safe havens, impose a partition line, or forcibly disarm all combatants even at the risk of having to

30. Robert A. Pape, *Bombing to Win: Air Power and Coercion in War* (Cornell University Press, 1996); Bradley Graham, "Joint Chiefs Doubted Air Strategy," *Washington Post*, April 5, 1999, p. A1.

31. See, for example, Stacy Sullivan, "Convince the KLA the War Is Over," *Washington Post*, June 6, 1999, p. B3.

32. See William Cohen, General Henry Shelton, and Major General Chuck Wald, Department of Defense news briefing, June 10, 1999; John A. Tirpak, "Victory in Kosovo," *Air Force Magazine* (July 1999), pp. 24–27; House Armed Services Committee, "Kosovo Update," June 25, 1999 (www.house.gov/hasc/kosovoupdate.htm [June 1999]). On the challenges of carrying out accurate bomb damage assessment, see Keaney and Cohen, *Gulf War Air Power Survey Summary Report*, pp. 55–144.

33. See Ivo Daalder and Michael O'Hanlon, "Unlearning the Lessons of Kosovo," *Foreign Policy*, no. 116 (Fall 1999), pp. 128–40.

fight them to do so. In military terms, as argued previously, technology seems unlikely to spark radical changes in infantry combat over the next two decades. And as discussed later in this chapter in regard to Kosovo, while airpower may improve significantly for certain kinds of warfare in the years ahead, its relevance to these types of missions is likely to remain limited.[34]

Technology might make a difference in other ways. For example, it could help make intervention forces lighter, thereby permitting them to be deployed very quickly to stop genocides or other severe forms of violence. Although many commentators overstated the difficulty that NATO would have had in deploying ground forces into Kosovo quickly, they were right that the alliance was not as capable in that regard as it should have been.[35] Trends in the technologies undergirding military vehicles, propulsion systems, and weapons, while short of revolutionary in promise (see chapter 6), might permit enough improvement to make a major difference in such limited missions. Some of what is needed to make forces lighter and more mobile— lighter tanks, more mine-resistant infantry vehicles, smaller sealift ships for use in less-developed harbors—is already technologically possible today (see chapter 7). Indeed, some key units and systems are already in the U.S. force structure, such as army air assault divisions and C-17 airlifters capable of carrying heavy equipment onto short runways. By taking advantage of these existing capabilities, as well as further improvements over time in the weight of armor and fuel efficiency of vehicles and lethality of munitions, it may be possible to rapidly deploy well-armed intervention forces in the future, at least for small missions. Specific new technologies, such as a large transport blimp, might make an important difference as well.[36]

Still, once troops arrive in a combat theater, they may need to fight in difficult infantry combat settings, and technology offers only modest hope that

34. Stephen John Stedman, "Spoiler Problems in Peace Processes," *International Security*, vol. 22 (Fall 1997), pp. 5–53; Roland Paris, "Peacebuilding and the Limits of Liberal Internationalism," *International Security*, vol. 22 (Fall 1997), pp. 54–89; Barry R. Posen, "Military Responses to Refugee Disasters," *International Security*, vol. 21 (Summer 1996), pp. 72–111; O'Hanlon, *Saving Lives with Force*, pp. 47–61.

35. John Barry, "Why Troops Take Time," *Newsweek*, April 26, 1999; Thomas E. Ricks, "Gung-Ho but Slow: Why the U.S. Army Is Ill-Equipped to Move Troops Quickly into Kosovo," *Wall Street Journal*, April 16, 1999, p. A1.

36. Thomas E. Ricks and Carla Anne Robbins, "NATO Plans Troops for Kosovo after Bombing," *Wall Street Journal*, May 5, 1999, p. A3; John Keegan, "Milosevic Keeps Allies Guessing as He Prepares to Play Waiting Game," *Daily Telegraph* (London), April 21, 1999 (Internet edition [October 1999]); Rachel Schmidt, *Moving U.S. Forces: Options for Strategic Mobility* (Congressional Budget Office, 1997), pp. 13, 80; Michael O'Hanlon, "Military Dimensions of a Ground War in Kosovo," Brookings Institution, April 26, 1999 (www.brook.edu/views/articles/ohanlon/1999unp.htm [September 1999]); Michael O'Hanlon, "Sins of Omission," *Washington Times*, May 24, 1999, p. A19.

an advanced military will be able to do so more effectively and safely in the decades ahead. On the whole, technology may offer some interesting improvements for peace operations and humanitarian interventions, but it is highly unlikely to change their basic character.

What about the post-conflict phases of such missions? Might technology make it possible, for example, to conduct peace operations in a place like Bosnia or Kosovo with fewer troops in the future? Sensors that listen for vehicles, radars that see through foliage, infrared cameras that allow humans to be distinguished from animals, and robotics that can attach themselves to vehicles or mark people will all reduce the need for human patrols along certain types of borders and zones of separation. These types of technologies, perhaps most notably microrobotic networks, could also help in the search for weapons stocks and other militarily significant assets. Such capabilities will be welcome.

However, the missions of monitoring and patroling do not generally determine troop requirements by themselves. Establishing and maintaining human intelligence networks typically require large numbers of troops and will continue to do so. Responding to violent outbursts or incursions will still require enough troops to conduct infantry combat operations, which, as noted previously, are unlikely to change greatly. Among other things, they will continue to rely on line-of-sight gunfire and on mechanized and helicopter transport technologies not notably faster than those that have been in use in recent decades. In fact, a survey of various types of stability operations since World War II shows no significant trend line in the numbers of troops that have been needed per 1,000 inhabitants. Required forces vary much more with the nature of the operation than with the state of communications, weapons, or transportation technologies.[37] This trend line is unlikely to change direction anytime soon. Again, the promise of technology for aiding humanitarian interventions and peace operations, while hardly negligible, is rather limited.

Case Study: Kosovo, 1999

Trends in airpower technology are likely to produce some important progress. Among the probable achievements, improved real-time communications systems will help a great deal. Although improvements have been occurring, today's weapons still do not always obtain targeting data quickly enough to destroy moving enemy assets before they can hide or take shel-

---

37. James T. Quinlivan, "Force Requirements in Stability Operations," *Parameters*, vol. 25 (Winter 1995–96), pp. 59–69.

ter.[38] NATO is also likely to be able to operate much more effectively in bad weather in the future.[39] Operation Allied Force against Serbia witnessed the use of munitions with somewhat better all-weather capabilities than those available during Desert Storm. Notable was the joint direct attack munition (JDAM), particularly when dropped by an aircraft with a precise radar like the radar on the B-2. (Other aircraft could carry equally good radar, although they did not in 1999.) Even these munitions were too inaccurate to target moving tanks, however. Submunitions with their own infrared, acoustic, and (someday perhaps) even radar sensors to automatically search for enemy armor will soon be able to attack below cloud cover without requiring aircraft to drop to low altitudes. Increased numbers of JSTARS aircraft and UAVs will provide the initial targeting information so that munitions can be dropped in the vicinities of targets at the correct times. The munitions may be severely challenged by complex terrain and some decoys and countermeasures, but in the open—where an aggressor may sometimes need to mass troops, as Serbia did against the KLA—they are likely to be effective.[40]

However, important limitations will remain in 2020. Enemy weapons hidden within buildings and forests are unlikely to be attacked successfully in most cases. The most important reason, as argued extensively in chapter 3, is that sensors will either not see targets in such places or have a hard time identifying and tracking them. Enemy supply lines are likely to be hard to shut down as well. Easily identified choke points and vulnerabilities will of course be targetable, even in bad weather. But assets such as small trucks using secondary roads and pontoon bridges will still be able to carry individual barrels of fuel or small stores of ammunition by avoiding main roads, otherwise minimizing their exposure to enemy sensors, and interspersing themselves within civilian traffic. There is some chance that new technologies, such as radio-frequency weapons, may allow airpower to shut down virtually all traffic into a province like Kosovo without causing civilian casualties; however, much stands in the way. For example, in wooded terrain, it may not be possible to get radio-frequency weapons near enough their targets to destroy the electronics. And at river and stream crossings, nonmotorized boats and ferries could move supplies without being vulnerable to such weapons.[41]

Finally and most fundamentally, even if NATO dominated the air in 2020 the way it did in 1999, units conducting ethnic cleansing with small and

38. Cook, "Serb Air War Changes Gear."

39. On the situation today, see Nick Cook, "NATO Battles against the Elements," *Jane's Defence Weekly*, April 21, 1999, p. 4.

40. William B. Scott, "Bad Weather No Deterrent for New Long-Range Weapons," *Aviation Week and Space Technology*, May 3, 1999, pp. 66–67.

41. O'Hanlon, "Military Dimensions of a Ground War in Kosovo."

medium weapons could still operate effectively unless challenged by a ground force of comparable strength. In previous guerrilla conflicts, a small rebel force immersed in a friendly population could resist much larger government forces. But the tactic of ethnic cleansing, as applied by Serbia against the Kosovar Albanians, puts the onus on the rebel force, which must try to defend its population rather than engage in hit-and-run warfare. In these conditions, a force like Serbia's—armed primarily with machine guns, mortars, and antiaircraft guns—can wreak havoc on a population like that of the Kosovar Albanians without possessing a huge overall force advantage against local insurgents. Since these types of weapons are easy to hide inside buildings and vehicles, they will remain hard for airpower to target in the future.[42]

In conclusion, for wars motivated by humanitarian concerns, technology is unlikely to change the basic situation NATO confronted in 1999: while an advanced military can use airpower to punish an aggressor, and limit its ability to conduct mechanized warfare, it cannot defeat small arms from long distances or stop the violence perpetrated by them.

Naval Combat

Some technologies relevant to naval combat are undergoing rapid change, while a number of others are not. On the whole, change in this area of warfare will probably be greater than in infantry combat—though less sweeping than for heavy ground combat.

The name of the game in naval warfare has long been to bring enemy ships under fire from the greatest distance possible. This trend accelerated with the advent of aircraft carriers. The vulnerability to attack of surface ships was exacerbated by cruise missiles in the second half of the cold war.[43] The advent of long-range Soviet Backfire bombers with advanced antiship cruise missiles brought this concern to its cold war peak in the late 1970s and early 1980s.[44] The sinking of two British ships by French Exocet missiles in the 1982 Falklands War, as well as the 1987 Iraqi Exocet attack against

42. Michael O'Hanlon, "Should Serbia Be Scared?" *New York Times*, March 23, 1999, p. A31; on counterinsurgency warfare, see Andrew F. Krepinevich Jr., *The Army and Vietnam* (Johns Hopkins University Press, 1986), pp. 7–16, 157–63.

43. See, for example, Paul H. Nitze, Leonard Sullivan Jr., and the Atlantic Council Working Group on Securing the Seas, *Securing the Seas: The Soviet Naval Challenge and Western Alliance Options* (Boulder, Colo.: Westview Press, 1979), p. 180.

44. On this matter, see Jerome Bracken and others, *Worldwide Barrier Air Defense of the Sea Lines of Communication*, IDA Report R-241 (Alexandria, Va.: Institute for Defense Analyses, 1978), pp. S-1 through S-30.

the U.S.S. *Stark*, provided vivid real-world demonstrations of these technical developments.[45] For more than half a century, the best—and perhaps the only—practical way to protect ships has been to keep the enemy from firing on them in the first place.

A natural corollary of these observations is that, in blue water hundreds of kilometers or more from land, today's U.S. Navy enjoys relative sanctuary. It can typically detect any enemy assets well before they reach firing range. In shallow waters and near shorelines, however, the situation is different, since U.S. ships could be ambushed. (Existing war plans assume that maritime pre-positioning vessels and fast sealift ships will be able to unload supplies in a place like Korea or Saudi Arabia within a few weeks of the beginning of a conflict. That may not be realistic, however, since it may take weeks to deploy antisubmarine and minesweeping forces to the theater and give them time to do their job.)[46]

Similar conclusions are true for undersea warfare. Tactical principles here are also simple: build the most sensitive sonars possible to detect other countries' submarines, understand the oceans as well as possible to take full advantage of all sonar data, and keep your own vessels as quiet and undetectable as possible. Again, this situation gives the United States great advantages in the open seas, where sonar propagation is fairly predictable and where other countries' vessels generally do not venture, but fewer advantages near shore, where sound propagation is complex and tactical distances are compressed.

Naval Combat in 2020

Ships themselves are likely to change only modestly over the next twenty years. They may become stealthier, but they will still be large targets difficult to camouflage when set against a body of relatively uniform reflectivity and temperature. They will become faster, but not radically so and—except for expensive nuclear-powered ships—will continue to require large amounts of petroleum-based fuel. The type of armor available to protect them will become lighter, though it will hardly be unobtrusive. Under the seas, submarines will become somewhat quieter, but still be detectable by sonar, particularly when moving at higher speeds or when illuminated by active sonar devices.

45. Lon Nordeen Jr., *Air Warfare in the Missile Age* (Washington, D.C.: Smithsonian Institution Press, 1985), pp. 201–03.

46. Owen Cote and Harvey Sapolsky, *Antisubmarine Warfare after the Cold War* (MIT Security Studies Program, 1997), p. 13; Schmidt, *Moving U.S. Forces*, p. 79; Frances Lussier, *Structuring the Active and Reserve Army for the 21st Century* (Congressional Budget Office, 1997), pp. 15–23.

Indeed, armor might have to become heavier in some cases, as in the battleships of decades gone by. The explosive power of conventional ordnance used against ships will probably not change greatly, but the likelihood of multiple hits against ships will increase due to the proliferation of cruise and antiship missiles, advanced sensors, quiet submarines, and advanced mines. This could greatly increase the incentives to protect them better; at present, modern ships can rarely sustain more than a few hits with large munitions before sinking.[47]

Antiship missiles will improve chiefly through advances in sensors, stealth, and maneuvering capability—and probably less so through propulsion systems, ranges, speeds, or explosive payloads. That means that ships operating hundreds of miles offshore will tend to remain reasonably safe, but those nearer shore will become increasingly vulnerable. Ship defenses will improve considerably; however, even if their terminal defenses prove partly effective, a large-scale saturation attack with cruise missiles would generally cause serious damage.

Improved diesel submarine technology will spread to more potential U.S. foes. Finding enemy submarines may thus require longer searches in the future, given that sonar technology does not appear to be on the threshold of major improvements (and that such submarines can remain remarkably quiet). If there is any good news here for the United States it is simply that these types of submarines will remain very constrained in their range and that they can generally still be located by active sonar. Moreover, that active sonar may, thanks to progress in robotics, increasingly be deployed on unmanned vehicles, reducing the danger to American sailors (since platforms using active sonar give away their own locations in trying to determine the enemy's).

U.S. minesweeping and mine-hunting capabilities are improving in a number of ways. Against a foe with current technology, that means the United States may soon be in a better position to breach shallow-water minefields. New mines may complicate this effort, however—particularly in an approach to enemy shores. For example, remote-controllable mines near shore might be activated only when enemy forces on land could visually verify the approach of a specific class of U.S. ship. Current minesweeping vessels that function by attempting to mimic high-value ships, thereby triggering the mines, may not work.

For all these reasons, approach toward enemy shores will be even harder in the future. The presumed success of that scenario against a foe with late

47. See *Jane's Naval Review 1987* (Surrey, England: Jane's Information Group, 1987), p. 124.

twentieth-century technologies hinged largely on being able to deny it advanced targeting information; for example, by using air assault forces to drive enemy infantry units out of the area where ships were to land and also by denying the adversary use of nearby airspace and seaspace. Against a moderately sophisticated foe, these tasks may get more difficult. Not only might such an enemy have UAVs, satellite systems, and other monitoring technologies to help it see U.S. forces coming, it would most likely have advanced cruise missiles and advanced remotely activated mines to profit from any targeting data that was available. Enemy forces could also use passive infrared detectors and their own eyes to fire on ships as they came within a few kilometers of shore or less. Countries attempting amphibious assaults may, to be sure, obtain somewhat better technology for doing so, such as the U.S. V-22 Osprey with its greater range and speed. But that platform does not offer radical improvements in survivability for assaulting forces.

Navies will find it increasingly dangerous to operate in close quarters like the Persian Gulf, although advanced supersonic cruise missiles do not appear to have been transferred there to date. In the early 1990s the United States strongly lobbied against Chinese plans to send such technologies to Iran. Even in the context of a strained dialogue on arms transfers, in light of U.S. aircraft sales to Taiwan and ongoing Chinese sales to Pakistan, Beijing realized how serious the United States was on this issue and desisted from high-technology missile transfers.[48] Had its premier antiship missiles been transferred, however, U.S. ships in the narrow Persian Gulf (just over 100 miles wide in places) and the more narrow Strait of Hormuz (about thirty miles wide) might have become vulnerable already—particularly if Iran had some dependable means of targeting them.

That distance of about 100 miles is also the width of the Taiwan Strait. However, by operating east of Taiwan in deep water, U.S. carriers may remain relatively safe in any future conflict with the People's Republic of China over Taiwan. Supersonic sea-skimming cruise missiles may become longer range, reaching beyond the tens of kilometers typical of today's systems. However, they may not be able to reach the multiple hundred-kilometer levels that would be needed without breakthroughs in propulsion technologies, which appear improbable even by 2020.[49]

This is not to trivialize the problem of defending U.S. ships in blue waters. A crafty foe may be able to think of creative ways to overcome the range

48. Robert S. Ross, "China," in Richard N. Haass, ed., *Economic Sanctions and American Diplomacy* (New York: Council on Foreign Relations, 1998), pp. 24–25.

49. *Jane's Weapon Systems 1988–89* (Surrey, England: Jane's Information Group, 1988), pp. 916–18.

limitations of many aircraft and missiles. For example, cruise missiles might be predeployed at sea along likely avenues of U.S. naval advance at a spacing of every fifty to 100 kilometers or so. They could be submerged until activated, then fly to a set of GPS coordinates corresponding to the ship's general location, where terminal homing sensors would kick in and guide the missiles to the ship.

In such settings, the United States might try to deny adversaries knowledge about the whereabouts of its carriers and other ships by shooting down imaging satellites and other reconnaissance devices and platforms. That raises the question of whether it should develop an antisatellite (ASAT) weapon in the near future. Doing so does not yet seem urgent. Moreover, there are broader strategic arguments against doing so at present, largely because of the state of Russia's early warning systems.[50] In the short term, moreover, ASATs would probably work against U.S. interests in conventional warfare. Most other countries are still many years away from developing substantial imaging satellite capabilities of their own but might be better able to develop crude antisatellite weapons if the United States leads the way politically and technically.[51] This situation could change over time, however, possibly making eventual U.S. acquisition of ASATs sensible.

What about so-called arsenal ships and mobile offshore bases? These systems, not yet built but clearly within technological reach over the time frame of interest, would have features to reduce their vulnerability. Since neither would be designed with speed in mind, both could be built with multiple hulls, armor, and other features intended to greatly reduce their vulnerability even if hit by a (conventional) explosive. In addition, the arsenal ship, in particular, could be built to have a small radar cross section and reduced visual, infrared, and acoustic signatures as well. Enemy detection ranges against this type of vessel could be reduced by at least a factor of three and

50. Bruce G. Blair, Harold A. Feiveson, and Frank N. von Hippel, "Taking Nuclear Weapons off Hair-Trigger Alert," *Scientific American*, November 1997, pp. 74–76.

51. This approach would probably require either directed-energy devices like lasers on the ground or the launch of "killer satellites." One might think the United States could flood low-earth orbits with debris (destroying all imaging satellites and other low-altitude devices, including its own), but that would require prohibitively large amounts of material. Assuming that one would populate most of the region within 1,000 kilometers of the surface of the earth with a layer of evenly distributed one-centimeter-diameter pellets, one would need about 1,000 pellets per square kilometer in this region. Large boosters might each be able to transport about 100,000 pellets (roughly ten tons of material), so each could cover a planar area of about 100 square kilometers. But there would be at least 10 million square kilometers of a cross-section of space to litter with these pellets, requiring 100,000 rocket flights. For a related article, see Nicholas L. Johnson, "Monitoring and Controlling Debris in Space," *Scientific American*, August 1998, pp. 62–67.

perhaps five to ten. Their resilience to direct hits might be better than for today's traditional vessels by a comparable percentage, meaning that their chances of being destroyed in certain tactical settings could be ten to 100 times less than destroyers, cruisers, frigates, amphibious ships, or carriers.

These vessels and mobile bases are slow, however, so they would be of optimal use only if predeployed in regions where conflict ensued. Their relative invulnerability would not extend to enemy attacks with nuclear weapons—hardly out of the question against a high-value military asset deployed in a place where collateral effects against civilian populations would be very limited even if the arsenal ship or mobile base were struck. Most important, they will be only as good as the munitions they carry. Arsenal ships and, eventually, mobile offshore bases, may well prove useful, but they will not necessarily represent radical or revolutionary advances.

Case Study: The Falklands War, 1982

In the 1982 Falklands/Malvinas War between Great Britain and Argentina, losses were substantial on both sides. Notably, the British lost six ships and about three dozen aircraft, including helicopters; Argentina lost a cruiser, a submarine, several small ships, and at least fifty aircraft. Britain suffered ship losses from Argentine Exocet missiles, from "dumb bombs" dropped directly on its ships' decks by Argentine aircraft that managed to penetrate defenses, and from the Argentine submarine.[52]

The U.S. Navy would not have suffered as many losses. Its carrier-based air defenses were much larger, measured by the number of aircraft involved, and much more effective than Britain's at that time (as they remain today). It is possible that Argentina's air force would not have penetrated U.S. carrier air defenses at all. The U.S. ability to find submarines might have been better than Britain's, too, but some losses would have been possible.

There are two main reasons for this: First, submarines are hard to find, particularly in confined waters, and they may wind up getting off a shot or two at even a good fleet before being found and sunk. Second, ships or helicopters approaching shore, as in any amphibious assault, are generally at risk due to mines and missiles; the troops they carry are at risk of ambush.

This situation will not be fundamentally different by 2020. Progress in sonar technology and other means of finding submarines is not, as noted, particularly rapid, though there is some hope that unmanned underwater vehicles using active sonar would permit enemy submarines to be found and

52. Nordeen, *Air Warfare in the Missile Age*, pp. 170–85.

destroyed without putting many U.S. personnel in harm's way. Regardless of the state of ship defenses, it has been true for decades that ships that can be brought under fire have a high chance of being sunk—if for no other reason than the fact that their defenses can be swamped by a sufficiently large attack. Moreover, the prevalence of antiship missiles and cruise missiles around the world is growing. Finally, as argued above, helicopters will remain at risk, as will the troops getting out of them; such forced-entry operations promise to remain very dangerous.[53]

Conclusion

New technology and associated tactical and operational innovations clearly have considerable potential to make important contributions to warfare by 2020. But it is equally clear that many fundamental limitations will be hard or impossible to overcome. That means combat will, in many ways, be similar in 2020 to its current nature. It also casts doubt on the hypothesis that a revolution in military affairs is under way and that a major transformation of the U.S. military is now required—or even prudent.

Infantry combat operations are likely to remain difficult and not particularly amenable to radical change in the first two decades of the century or so. A number of specifics about the way they are conducted will change, but the changes will work both for and against U.S. troops, without a strong bias toward one way or the other. Sensors will not improve so much as to make ambushes impossible.

Strategic and battlefield mobility and protection will change only modestly. Basic propulsion systems and fuel requirements will improve gradually, but at evolutionary—rather than revolutionary—rates. Most airplane speeds, payloads, and costs will not change drastically. Space launch will get cheaper but remain relatively expensive and difficult. Movement on the battlefield will continue to be provided by land vehicles moving no more than several tens of miles per hour or aircraft with speeds comparable to today's. Helicopters and transport aircraft will remain quite vulnerable when flying near the earth's surface.

Exposed armor will undoubtedly become somewhat more vulnerable over time. Large mechanized formations unambiguously poised near a potential victim's borders will be even more vulnerable in the future than they are

53. See, for example, Nitze, Sullivan, and the Atlantic Council Working Group, *Securing the Seas*, p. 180.

now, particularly if substantial U.S. forces are nearby. But when armor is not exposed, it will still be hard to attack.

As for war at sea, on the whole, high-value naval assets approaching prepared positions are likely to become more vulnerable in the twenty-first century.[54] The increased vulnerability is likely to be a result of enemy mines and submarines, which will most likely improve at least as fast as mine warfare techniques and sonar. Ships may become more vulnerable at greater distances from shore, though such trends are less clear and probably less notable.

Some technological concepts have promise but are far from being realized. They include directed-energy weapons, acoustic weapons, and kinetic energy weapons based on land or perhaps in space.

Consider again *Joint Vision 2010* and its emphasis on dominant maneuver, precision strike from long distances, full-dimensional protection, and focused logistics, all under the umbrella of overwhelming information superiority. Although not precisely defined in the document, the concepts of dominant maneuver and full-dimensional protection imply near-term breakthroughs in areas like missile defense and lighter armor that, as noted previously, do not appear to be on the immediate horizon. Given that most major military systems require more than a decade to acquire, the document's vision already appears doomed to fall far short of its goals. In particular, the following passage seems almost wildly optimistic and premature: "By 2010, we should be able to change how we conduct the most intense joint operations. Instead of relying on massed forces and sequential operations, we will achieve massed effects in other ways. Information superiority and advances in technology will enable us to achieve desired effects through the tailored application of joint combat power."[55] There is a kernel of truth in this: it is true that information networks will improve in coming years, and that homing munitions will become more effective as well. However, the suggestion that firepower will become far more precise by 2010 than it was in the mid-1990s, when these words were written, is unsubstantiated by the basic thrust of technological change.

Recall, too, the air force claim in 1997 that "in the first quarter of the 21st century you will be able to find, fix or track, and target—in near real-time—anything of consequence that moves upon or is located on the face

54. See Vickers and Martinage, *The Military Revolution and Intrastate Conflict*, pp. 15–19; Andrew F. Krepinevich Jr., *The Conflict Environment of 2016: A Scenario-Based Approach* (Washington, D.C.: Center for Strategic and Budgetary Assessments, 1996), pp. 7–15.

55. John M. Shalikashvili, *Joint Vision 2010* (Department of Defense, 1995), pp. 17–27, especially pp. 17 and 19.

of the Earth."[56] That may be true for easily recognizable military objects like tanks and ships (provided they can be distinguished from allied and civilian assets). It is unlikely to be true (particularly at standoff distances) for small arms, enemy soldiers interspersed among a background population or in a heavy forest, mortars and antitank and antiaircraft missiles hidden in trucks or caves, and properly secured weapons of mass destruction.[57] The air force was too sanguine on this particular assessment.

The grandiose nature of such claims and goals, and the absence of concrete technological means of backing them up, are somewhat reminiscent of the most extreme aspirations of the promoters of the 1980s Strategic Defense Initiative (SDI). There may have been legitimate grounds for supporting that program for its technical research agenda, and an argument to be made that it hastened the demise of the communist bloc by starkly reminding the Soviet Union that it could not compete technologically with the West, but the initiative's goal of being able to bear the brunt of a major Soviet nuclear strike was highly unrealistic. The SDI experience should remind participants in the current RMA debate that willpower alone is not sufficient to create technological solutions to problems—particularly those that an adversary has an interest in opposing. As Ben Rich, the former director of Lockheed's Skunk Works and father of the F-117 fighter, wrote in criticizing SDI and offering guidance about how to think about future military innovation more generally, we will waste money in defense unless "the new technology we are attempting to create is really within practical reach of our current abilities and achievable with reasonable expenditures."[58]

It is doubtful that a military revolution can be driven primarily by breakthroughs in computer technology. However fast and efficiently an advanced electronic grid can process and transmit data, that data will be useful only when it is reliable and accurate. Computers and communications cannot compensate for many of the limitations on sensors that are likely to continue to inhibit performance in the foreseeable future. Without good data, the old adage about computers applies: garbage in, garbage out. To take one example, as Lawrence Freedman writes, despite their overwhelming fire-

56. See General Ronald R. Fogleman, Chief of Staff, U.S. Air Force, statement before the House National Security Committee, 105 Cong. 1 sess., May 22, 1997.

57. Martin C. Libicki, "DBK and Its Consequences," in Stuart E. Johnson and Martin C. Libicki, eds., *Dominant Battlespace Knowledge,* rev. ed. (National Defense University, 1996), pp. 44–46; Alan Vick and others, *Enhancing Air Power's Contribution against Light Infantry Targets* (Santa Monica, Calif.: RAND, 1996), pp. 13–30.

58. Ben R. Rich and Leo Janos, *Skunk Works* (Boston: Little, Brown, 1994), p. 337.

power advantage over Iraq at the time of the standoff over U.N. weapons inspections in 1997–98, the allies "lacked 'dominant situational awareness'" about where to find Iraq's chemical and biological capabilities—something that improved computer networking and faster processing of available data will not be able to change as long as the raw data itself is of limited value.[59]

Improvements in military technology will certainly be impressive in the next twenty years and will certainly have important implications for warfighting. But, as the director of the Defense Intelligence Agency has recently argued, even if future warfare is characterized by major changes in technology, tactics, and strategy at some levels, it may change much less in other important milieus and settings.[60] Unfortunately, those latter settings—largely the realms of urban and infantry combat—may well be the places where the United States faces its greatest security challenges in the decades ahead.

One cannot definitively disprove the RMA hypothesis, especially for high-intensity armored warfare. Modern electronics and computing capabilities do offer the promise of new patterns of warfare through much more rapid information processing and dissemination. But the counterarguments to the RMA hypothesis are legion, and the case for basing a radical transformation of the U.S. military on the contemporary RMA notion is unpersuasive at present.

59. Freedman, *The Revolution in Strategic Affairs*, p. 47.
60. Patrick M. Hughes, "Global Threats and Challenges: The Decades Ahead," Defense Intelligence Agency, Washington, D.C., 1999, p. 15.

CHAPTER SIX

The RMA Hypothesis
and U.S. Security Policy

Proponents of a contemporary revolution in military affairs (RMA) have not confined their prognostications to military force planning and budgeting. The more expansive and ambitious RMA schools of thought see sweeping implications for U.S. security policy across a wide range of issues. Three of their most important claims concern the future need for overseas U.S. bases, the future nature of multinational military operations, and the possibility of eliminating nuclear weapons worldwide as well as reducing the vulnerability of U.S. forces to other types of weapons of mass destruction. Many RMA advocates believe firmly that the need for overseas bases and military deployments will decline radically if the United States takes proper advantage of what current and next-generation technologies can offer. Less optimistically, many also believe that the purported RMA will be an American phenomenon, leaving U.S. allies in the high-tech dust and making it increasingly difficult to conduct combined military operations on a multinational basis. Finally, an admittedly smaller number of RMA proponents hope that trends in sensor technology, information processing, and verification will make it possible to eliminate nuclear weapons from the face of the earth within decades, as well as greatly reduce the military's vulnerabilities to chemical and biological weapons. This chapter argues that all three of these claims are essentially wrong for the foreseeable future.

Forward Presence and Overseas Basing

One argument made by RMA proponents that has broad foreign policy implications is that technology may offer a way to scale back U.S. global military presence substantially—if not right away, then certainly within the first decade or two of the twenty-first century.[1] It is a refrain heard from bomber advocates, the U.S. Air Force, and some regional security experts who have begun trying to apply the RMA concept to theaters such as the Asia-Pacific.[2] The specific mission under consideration might involve stopping another Iraqi or North Korean thrust southward against U.S. allies. It might also involve a Russian move against the Baltic states or the oil reserves of the Caucasus region, or an attempted invasion of Taiwan by the PRC. Is this prediction of a drastically declining need for overseas bases accurate?[3] The short answer is no. Through 2020 or thereabouts—the time frame of this study and the planning horizon of military experts today—most major overseas operations will continue to require good access to bases near a conflict. It may be possible to reduce force numbers modestly without sacrificing capability. It may also be possible to reduce overseas military presence and operations in peacetime in some circumstances, maintaining plans to build up forces in crisis or wartime if necessary. But the United States will not be able to conduct major combat operations from its own territory.

Although reduced more than 50 percent from cold war levels, U.S. forces based or deployed abroad number 250,000 uniformed personnel. Forward-deployed forces now constitute just over one-sixth of the 1.4 million active-duty force—down from one-quarter of the total in cold war times—but many of the remaining deployments are particularly demanding. Troops in Korea, Bosnia, and Kosovo, most marines on Okinawa, air force pilots in Saudi Arabia, and, of course, navy sailors and marines at sea are unescorted by

---

1. Michael O'Hanlon, "Can High Technology Bring U.S. Troops Home?" *Foreign Policy*, no. 113 (Winter 1998–99), pp. 72–86.

2. See Rebecca Grant, ed., *Origins of the Deep Attack Weapons Mix Study* (Arlington, Va.: IRIS Independent Research, 1997), p. 16; James R. Blaker, "Understanding the Revolution in Military Affairs: A Guide to America's 21st Century Defense," Defense Working Paper 3 (Washington, D.C.: Progressive Policy Institute, January 1997), pp. 24–26; Chrisopher J. Bowie, *Untying the Bloody Scarf* (Arlington, Va.: IRIS Independent Research, 1998), p. 17; Charles M. Perry, Robert L. Pfaltzgraff Jr., and Joseph C. Conway, *Long-Range Bombers and the Role of Airpower in the New Century* (Cambridge, Mass.: Institute for Foreign Policy Analysis, 1995), pp. ix–xxii; Paul Dibb, "The Revolution in Military Affairs and Asian Security," *Survival*, vol. 39 (Winter 1997–98), p. 106.

3. A similar version of this section appears in O'Hanlon, "Can High Technology Bring U.S. Troops Home?"

their families. Forces on Okinawa are much less welcome than they once were; air force units in Saudi Arabia suffered a deadly attack against their barracks in 1996 and remain a focal point for terrorists; and the tragic accident in which a marine aircraft sent twenty tourists plummeting to their deaths in 1998 strained the U.S.-Italy relationship as well.

How much nicer it would be if U.S. troops could stay at home until called upon in a crisis or conflict. According to some RMA believers, that should be possible in the near future. Under their vision, U.S. forces could lash out rapidly, intercontinentally, and lethally from U.S. bases with spacepower, long-range airpower, and other twenty-first-century gadgetry. Although it did not go quite that far, the 1997 National Defense Panel (NDP) did offer a similar vision, arguing for a future military capable of "inserting and extracting forces in the absence of forward bases" and projecting "significant power from forward deployed areas, as well as the United States, within hours or days rather than months."[4]

Alas, this image of future warfare succumbs to the most unrealistic aspects of the RMA hypothesis, while failing to pay heed to real-world technical constraints that are likely to remain with us for decades. Among the enduring realities it overlooks are the following:

—Most combat airplanes will remain short range, given the realities of aerodynamics and engine technology.

—Ground forces may become somewhat lighter, but they will hardly be light; they will continue to depend on motorized and armored vehicles that require large amounts of fuel and other logistics support.

—Transport airplanes will not have enough payload or speed to deploy large armored forces themselves.

—Long-range sensors will remain of limited value against a number of types of militarily relevant assets, including most weapons of mass destruction and small arms.

—Ships will get faster but remain fairly slow, requiring many days or weeks to cross oceans.

—It will remain very difficult to seize ports, airfields, and other fixed infrastructure held by an enemy.

Consider first the air superiority mission. Even if bombers operating out of U.S. bases are someday able to provide much of the air-to-ground punch in future conflicts, other aircraft will be needed to patrol the skies. More-

4. National Defense Panel, *Transforming Defense: National Security in the 21st Century* (Arlington, Va.: December 1997), pp. 35, 46.

over, those aircraft will need dependable bases in the actual theater of combat, either afloat or (better yet, to the extent possible) ashore.

The kinds of new aerial vehicles being researched today will not allow the United States to establish and maintain air superiority out of bases on its own territory. Unmanned aerial vehicles (UAVs) are primarily short range; jet engine technology is not advancing enough to make intercontinental flights quick or fuel efficient; extremely fast hypersonic vehicles, if they can be built at all, will be very expensive and specialized in their purposes.

The number of advanced aircraft needed to conduct the patrols, fighter sweeps, and other key operations inherent in establishing and maintaining air superiority in a future overseas war is open to some debate. In fact, in the 1997 Quadrennial Defense Review, the Pentagon appears to have revised its previous estimate downward, determining that perhaps 1.5 wing equivalents of F-22 aircraft (roughly 100 to 125 operational aircraft) would be adequate given the attributes of that jet.[5] However, that is still a prohibitively large number to maintain from U.S. bases, and it does not even count the support aircraft that would also be needed. It would be physically impossible for a pilot to conduct such missions with the F-22 or other fighter jets, given the limits of fatigue. Even if future fighter aircraft were unmanned, a slight technical possibility by 2020, placing twelve to twenty-four hours of (roundtrip) strain on an aircraft's airframe and engines and wasting huge amounts of fuel every sortie would be impractical.

Why will air superiority still be essential in 2010, 2020, and beyond? For one thing, it will be needed just to make good use of stealthy ground-attack aircraft like the B-2 bomber, which could operate from American soil if necessary. The B-2 is not invisible or invulnerable; it is just hard to see on radar.[6] As such, it can be visually detected in sunlight. An enemy worried principally about the B-2 could expose its own forces only during the daytime, digging in for cover at night. Given the B-2's inability to outfly even unsophisticated fighters, it would fare poorly under these circumstances if unescorted.

In fact, B-2 bombers might even need fighter escort at night against a moderately sophisticated enemy. Even in the 1999 NATO war against Ser-

5. William S. Cohen, *Report of the Quadrennial Defense Review* (Department of Defense, May 1997), p. 45.

6. It may become easier to see in the future than it is now, moreover. If enemies can develop effective low-frequency radars or use their commercial low-frequency radio transmissions as a "natural" radar beacon, they may be able to improve detection ranges. See David A. Fulghum, "Passive System Hints at Stealth Detection," *Aviation Week and Space Technology*, November 30, 1998, p. 70.

bia, the air force generally sent them into battle with radar jammers—despite the fact that the United States and NATO owned the sky over Serbia. The B-2 may become more vulnerable in the future, depending on what happens to radar technology. It may become more visible against bistatic radar, or perhaps a longer-wavelength or broader-bandwidth device, over time. It is also possible that B-2s could give their own locations away when they used their radars to search for targets. Their low-energy radars are designed to make it hard for an enemy to track them in this way. But if an enemy had numerous radar sensors with real-time data links between them, it might locate the B-2 through a triangulation technique (by measuring the times of arrival of a given search beam at different points and inferring the aircraft's position from the differences in those times).[7]

Could the United States rely on missiles launched from arsenal ships instead of B-2s? These futuristic vessels would have little superstructure to give away their locations, require only small crews, and carry perhaps 500 medium-range missiles. The missiles could each hold ten or more submunitions to destroy armored targets as well as other enemy assets.

Arsenal ships may or may not prove as good as now claimed by their proponents. For one thing, they would need targeting data to know where to fire their missiles—quite possibly requiring short-range aircraft or UAVs as a result. In addition, they can be only as effective as the munitions they carry, and those munitions would have to attack targets that would surely use dispersal, decoys, camouflage, jammers, and other techniques to defend themselves.

To be effective in the early days or weeks of battle, arsenal ships would have to already be located in the right part of the world when conflict broke out. The United States might also have to respond quickly, when enemy forces were most exposed during their acute attack phase. Any indecision over how to react could waste whatever window of opportunity the arsenal ships might provide. Yet policymakers usually hesitate, even when friendly countries are attacked; they may need time to prepare the nation (and themselves) for combat or to determine what caused a war and how to apportion blame for its initiation. Nothing about future military technology will make these types of decisions, which are intrinsically political and strategic, much easier. Putting a great deal of stock in arsenal ships thus

7. See Robert Wall, "U.S. Seeks Innovative Ways to Counter Air Defenses," *Aviation Week and Space Technology*, July 20, 1998, p. 54; Beth M. Kaspar, "Advanced Tactical Targeting Technology," paper prepared for DARPA Tech '97 Systems and Technology Symposium, Kansas City, Mo., September 1997.

seems imprudent—even if it turns out to be wise to buy a number of them eventually.

The need for short-range combat aircraft, and hence significant numbers of well-protected forward bases, seems likely to endure. Air superiority will also be needed to protect any U.S. and allied ground operation. It is easy to see why, even in an era of faster and more "brilliant" munitions and computer networks, ground forces will still be important in war. For one thing, as argued previously, weapons such as B-2s and arsenal ships may not prove effective in the early phases of a war, whether for military or political reasons, and an ally may be defeated as a result. That could make it necessary for U.S. forces to help liberate its ally through infantry and urban combat. The United States and its allies may also need to use ground troops to overthrow an extremist enemy regime that is developing nuclear weapons, committing genocide, or otherwise causing an acute international crisis.

To accomplish these things, not only tactical airpower, but also ground troops, will remain essential. Long-range reconnaissance assets will continue to have trouble finding and attacking enemy forces that have hunkered down or mixed in with civilian populations. They will be especially challenged to locate critical military assets like mortars, man-portable antitank and anti-air weapons, rocket-propelled grenades, soldiers, underground bunkers and communications facilities, and well-concealed weapons of mass destruction. Significant numbers of forces on the ground using short-range sensors will be needed to find and attack the enemy in these situations.

The reasons for this are not hard to understand. As discussed in chapter 3, all sensors have limitations, and to a large extent they are limitations imposed by the basic, immutable laws of physics, not simply by the temporary state of current technology. Visible-light and infrared detectors cannot see through heavy clouds. Neither these types of visible and near-visible sensors nor radars can penetrate metal, water, or most kinds of soil very far. Most X-ray and particle-beam sensors have very short ranges. Some sensor improvements are on the way, but they will probably not make it possible to see through tree trunks, discriminate camouflaged heavy equipment from trucks and other civilian vehicles, or see inside any of those vehicles.

As discussed, some new sensors and weapons are being designed for close-in infantry operations. Robotics may allow individual soldiers to investigate areas behind or within nearby buildings without exposing themselves to fire; soldiers may then be able to attack enemy forces hidden in such places with miniaturized maneuvering mortar rounds of one type or another. Acoustic sensors may allow troops to pick up the precise location of an enemy sniper from the report of his weapon and high-frequency microwave radars may allow troops to "see" through a nearby wall.

However, these sensors and weapons will remain short range: they will not be able to peer or fire through thick and multiple layers of concrete or soil; they may or may not work in thick forests; and they will not have magical powers to tell good guys from bad guys or divine the unspoken intents and plans of the enemy. They are impressive concepts, but perhaps no more so than the automatic rifles, man-portable surface-to-air missiles, night-vision technology, and helicopters that infantry forces have already acquired in the last half-century.

Future ground forces may be able to carry less heavy weaponry with them in the future, calling in precision fire support when needed, as a 1996 Defense Science Board summer study suggested.[8] However, significant constraints will remain. Scout teams may be vulnerable and hard to resupply. Even hypersonic missiles would generally take tens of seconds or minutes to deliver their ordnance—a time lag that will not always be acceptable against enemy forces that are able to move and take shelter. Moreover, even if some missiles could loiter over a battlefield so as to be able to deliver submunitions at targets more quickly, these capabilities would be of limited utility in heavy forest or urban environments. Thus ground forces will need to carry a certain amount of weaponry with them. They will also still need a certain amount of armor and organic mobility for self-defense and offensive fighting purposes. That means they will remain fairly large and heavy, even if somewhat less so than today.

Due to their weight, ground forces will continue to rely on sealift—meaning they will also be relatively slow to deploy. Even medium-weight ground forces are too heavy to fly anywhere in large numbers within reasonable time periods. Soldiers will still need water to drink, fuel to power their vehicles, and relatively heavy armor to provide protection. Advances in engine efficiency, armor, and other technologies will improve things, but only incrementally.[9] Weapons will become more accurate, lessening demands on ammunition stocks. But even this development needs to be seen in perspective. Even today ammunition does not become a significant fraction of transported supplies until combat has been under way for weeks. So cut-

8. Trevor N. Dupuy, *Attrition: Forecasting Battle Casualties and Equipment Losses in Modern War* (Fairfax, Va.: Hero Books, 1990), pp. 28–29.

9. For a similar assessment, see Defense Science Board 1998 Summer Study Task Force, *Joint Operations Superiority in the 21st Century* (Department of Defense, 1998), p. N-1.

ting back requirements will not solve major problems; improvements will be of degree, not of kind.[10]

Today it would require 3,000 strategic transport flights to airlift a heavy division with all of its vehicles, and more than 15,000 for a force of four heavy divisions, including its corps-level support assets.[11] It will still probably take 10,000 in twenty years. Even assuming a doubling of the strategic airlift fleet's size—with associated costs of about $100 billion—at least two months' time would be needed simply to deploy the initial forces, assuming assured access to airfields throughout, followed by a comparable effort to resupply forces thereafter. This is not a practical way to move armor, particularly in large amounts.

Sealift technologies will change only gradually. Even if major breakthroughs are achieved, ships will do well to sail at forty or fifty knots in contrast to today's speeds of twenty to thirty knots. That means that the NDP's goal of being able to deploy significant forces "within hours or days rather than months" is generally unrealistic. Forces can be made easier to load and unload, and easier to airlift in modest amounts, but it will still take time to move them—perhaps not many months, but certainly several weeks. Achieving such a significant improvement in rapid mobility would be militarily important. But it would not necessarily be revolutionary in that an enemy would still have a good deal of time to attack and seize territory before the United States could respond in force.

To be sure, there are potential scenarios in which technological developments likely to occur over the next twenty years, together with better uses of existing technology, could make a critical difference in how U.S. forces would deploy overseas. One noteworthy set of examples concerns humanitarian interventions. In some such cases, such as Rwanda during the genocide of 1994, Western infantry forces that could have been (but were not) quickly airlifted to the scene had plenty of firepower to dominate battlefields right away. In cases such as NATO's war against Serbia in 1999, however, the potential opponents could be strong enough to challenge current U.S. capabilities. Although, in my judgment, the United States and its allies could nonetheless have mounted a decisive ground invasion of Kosovo after

10. On ammunition usage rates, see Major John E. Edwards, *Combat Service Support Guide*, 2d ed. (Harrisburg, Pa.: Stackpole Books, 1993), pp. 152–53; on the relative share of army weight for a large combined-forces deployment (predicted to be roughly 75 percent), see Rachel Schmidt, *Moving U.S. Forces: Options for Strategic Mobility* (Congressional Budget Office, 1997), p. 7; on the weight of the Desert Storm deployment, see Department of Defense, *Conduct of the Persian Gulf War: Final Report to Congress* (April 1992), pp. F-24 through F-25.
    11. Schmidt, *Moving U.S. Forces*, p. 80.

six to eight weeks of preparation, doing so would have stretched the capabilities of NATO's long-range airlift forces, helicopter-mobile units, and marine corps amphibious shipping. By contrast, a future force made up of more fuel-efficient and lighter vehicles and more lethal weaponry might make such an operation possible after only a couple weeks of deploying forces abroad. Future developments in transport technology, such as the mega-blimp discussed in previous chapters, could help speed things up as well. Looking across a broader range of possible military scenarios, however, including much more demanding missions than a possible ground invasion of Kosovo, trends in power projection technologies do not appear to offer revolutionary improvements. It is doubtful that, by 2020 or thereabouts, the United States and major allies will be able to conduct extremely rapid power projection against most opponents directly from home bases.

Taken together, these conclusions mean that the United States should continue basing military forces overseas unless allies are capable of providing for their own security. At a minimum, enough U.S. forces to secure airfields and ports will be desirable, to the extent that allies in the region cannot be counted on to do this job. The alternative is to lose the initial battles and perhaps, ultimately, be forced to fight ashore. Forced-entry operations will only become more dangerous with time, however. Enemies will retain a range of weapons from mines to antiship cruise missiles to advanced antiaircraft missiles to submarines, as well as chemical and biological agents, to protect the seacoast and land they control.

None of this analysis is to argue that forward bases will go unchallenged by future adversaries. Indeed, the United States will have to worry a great deal about the vulnerability of overseas bases and try to disperse, harden, and defend bases in increasingly sophisticated ways.[12] However, the idea of returning forces to the United States and becoming in effect an "isolationist global power" is taking the RMA hypothesis too far.[13]

These stubborn technical facts about the enduring need for U.S. foreign military presence may disappoint RMA proponents, but they should actually be welcomed by the rest of us. By routinely operating abroad, U.S. armed forces can develop professional contacts with other militaries, ensure the interoperability of U.S. and allied equipment, work out plans for various

12. See Dr. John S. Foster, "Defense Science Board Summer Study: Investments for 21st Century Military Superiority," briefing papers, November 1995; Andrew F. Krepinevich Jr., "Transforming the American Military" (Washington, D.C.: Center for Strategic and Budgetary Assessments, September 1997), p. 7; Perry, Pfaltzgraff, and Conway, *Long-Range Bombers and the Role of Airpower in the New Century*, pp. 23–31.

13. I thank Mark Strauss of *Foreign Policy* magazine for coining that phrase.

scenarios together, and train to comparable standards. Taken together, these types of efforts greatly increase the odds that future multilateral military operations will proceed smoothly and successfully.

There are broader strategic benefits to forward presence as well. Deterrence tends to be strongest when a potential aggressor knows that U.S. forces would suffer casualties in any attack it might attempt. Forward basing and deployments had a strong track record of achieving deterrence in Europe during the cold war and have been similarly effective in Kuwait and Korea since U.S. forces were permanently stationed there. They also help demonstrate that the United States is committed to freedom of navigation in places like the Persian Gulf and South China Sea. By contrast, paper treaties not backed up by manifestations of military commitment are less credible, because historically they have not always been honored by their signatories in times of crisis.

This is not to argue that U.S. forward presence overseas will have to remain at its current level of 250,000 troops forever. But the impetus to any further reductions is more likely to be strategic than technological. An eventual end to the Korean military standoff, for example, could motivate a cutback of 10,000 to 20,000 U.S. troops from the peninsula. U.S. forces in Japan or Germany might someday be reduced by a comparable amount to reduce strains on troops and on host countries. Equipment could still be stored on their territories, just as it was during the cold war in Germany (and as it is today on a smaller scale in places like Guam, Diego Garcia, and Kuwait). Troops could then quickly fly back to operate the prestationed equipment should that be needed.

To be sure, future technological innovations also could contribute to reductions in U.S. military presence abroad. For example, if the typical number of aircraft in a fighter wing declines by 25 percent, the number of people required for the wing might decline more or less proportionately. (Indeed, army heavy divisions are already slated for roughly a 15 percent thinning.) Marine combat battalions may become considerably smaller in the future, navy surface combatants will be more computerized and require smaller crews, and UAVs might ultimately reduce the size of an air wing by eliminating the need for pilots (typically 10 percent of a wing's personnel).[14] These changes will be modest and incremental, however. Whether an RMA takes

---

14. For similar views, as well as discussion of various scenarios by which U.S. force presence abroad could also be changed for geostrategic reasons, see Richard L. Kugler, *Changes Ahead: Future Directions for the U.S. Overseas Military Presence* (Santa Monica, Calif.: RAND, 1998), pp. 120–23, 139–69.

place or not, the basic need for overseas facilities and deployments will remain.

Military Innovation and Allied Operations

Since 1990 the United States has often been described as the sole surviving superpower. This nickname is particularly apt in the realm of conventional military operations. The degree of U.S. supremacy is remarkable—not just one order above any other countries but two discernible steps. Not only does the United States spend five times more on defense than any other country, it also spends its money better, producing a far more effective return on its investment than any of its major NATO allies or Japan with the partial exception of Great Britain. Among the capabilities it possesses that most other countries do not even own in proportionate terms are: long-range strategic transport, mobile logistics, advanced precision-guided weaponry, stealth technology, and global surveillance and communications systems.[15]

One might think that NATO Europe, which in aggregate spends two-thirds as much on defense as the United States and sustains even more people under arms, might have at least 50 percent of the capability of U.S. armed forces.[16] Although it may attain or even exceed this figure on its own European territory, it does not come close beyond NATO's borders. In fact, beyond their own continent, European countries in aggregate would do well to keep up with the U.S. Marine Corps, which makes up just 12 percent of total American military strength.[17]

This situation, even if it is convenient for America's allies and desirable to some in Washington who enjoy the influence it accords the United States, is very unhealthy for the Western alliance. Even if some policymakers might like to preserve this global correlation of forces and use it to establish a form of American hegemony, most of the American people and most of the rest

15. For a similar treatment, see Michael O'Hanlon, "Military Innovation and Allied Operations," *National Security Studies Quarterly*, vol. 5 (Spring 1999), pp. 75–82.

16. For an argument that it should have such a capability, see David Gompert and Richard Kugler, "Free-Rider Redux," *Foreign Affairs*, vol. 74 (January–February 1995), pp. 7–12. For data on its capabilities, see International Institute for Strategic Studies, *The Military Balance 1997/98* (Oxford University Press, 1997), p. 293.

17. See John E. Peters and Howard Deshong, *Out of Area or out of Reach? European Military Support for Operations in Southwest Asia* (Santa Monica, Calif.: RAND, 1995), pp. 77–120; Michael O'Hanlon, "Transforming NATO: The Role of European Forces," *Survival*, vol. 39 (Autumn 1997), pp. 5–15.

of the world seem unlikely to tolerate such an approach.[18] First, it would reduce the prospects that U.S.-led operations will receive enough international backing to gain strong legitimacy in the eyes of the global community. Second, it could make it harder to secure access to overseas bases (which, as argued, will remain very important). Last, it would risk a situation in which, after suffering substantial losses in a future war in defense of common Western interests, the American people become fed up with allies they see as freeloaders and pressure their leaders to abdicate the U.S. leadership role—perhaps even pulling out of some major alliances. Such a backlash may not even require a major war. If the United States continues to spend a considerably higher percent of its gross domestic product (GDP) on defense than most of its allies, and experiences a protracted economic downturn, isolationist sentiments could arise even in peacetime.

If this happens, both U.S. global leadership and the Western alliance system could be seriously weakened. Some may believe the twenty-first-century world will be a sufficiently safe place that this would not matter. Yet few analysts and statesmen expected the twentieth century to be particularly conflict-ridden or bloody, and they were proven wrong. International relations have tended to be very dangerous when the world lacked a clear security anchor in the form of one or (at most) two strong powers or alliance systems. It would be irresponsible to base policy on optimistic assumptions about the obsolescence of war.[19]

The RMA debate risks making this problem worse. Rather than seeking to redress these imbalances in deployable military power, allies may decide it is pointless to try to keep up with the United States. For example, in the words of a Dutch general in 1997, "looking at the development of Army XXI in the U.S., I wonder whether the other NATO countries are able to keep up with our ally."[20] Likewise, a British officer stated that "digitization might make coalition warfare all the more difficult."[21] Lawrence Freedman worries about the possibility of a two-tier alliance in which the United States

18. Richard N. Haass, *The Reluctant Sheriff* (New York: Council on Foreign Relations, 1997), pp. 54–55.

19. See Kenneth N. Waltz, *Theory of International Politics* (Reading, Mass.: Addison-Wesley, 1979), pp. 129–93; Robert Gilpin, *War and Change in World Politics* (Cambridge University Press, 1981), pp. 231–44; Aaron L. Friedberg, "Ripe for Rivalry: Prospects for Peace in a Multipolar Asia," *International Security*, vol. 18 (Winter 1993–94), pp. 5–10; Donald Kagan, "Locarno's Lessons for NATO," *Wall Street Journal*, October 28, 1997, p. A22.

20. "US Forces' Digital Revolution Threat to Interoperability," *Jane's Defence Weekly*, June 11, 1997, p. 17.

21. "Task Force XXI Portends Future Interoperability Problems with Allies," *Inside the Pentagon*, April 10, 1997, p. 24.

provides high-tech capabilities while its allies perform the more mundane and dangerous old-fashioned work.[22]

American officials share some of these concerns. On a trip to Australia in 1998, Secretary of Defense William Cohen suggested that at least that particular U.S. ally was in danger of falling so far behind American forces technologically as to reduce its ability to participate usefully in coalition operations.[23] Under Secretary of Defense Jacques Gansler and Joint Chiefs Chairman Henry Shelton commissioned a Defense Science Board study on the topic of U.S.-led coalition operations out of concern that advanced technologies could make them more difficult.[24] In Operation Allied Force against Serbia in 1999, the U.S. military conducted an overwhelming majority of the precision attacks, leading the U.S. Air Force general who ran NATO's air war to worry that most U.S. allies were at risk of becoming junior partners in the alliance.[25]

There is no solid military or economic reason that advanced technology should make coalition operations more difficult, however. If one discards the sweeping theoretical language that has characterized much of the RMA debate to date and focuses on more concrete military issues and technologies, it is possible to see why.

Advanced munitions, computers, avionics, and communications—the areas where military technology is changing fastest—do not cost so much as to be out of financial reach of major allies. Future battlefield integration and interoperability will depend on having computers and communications systems that can talk to each other across national lines. That is admittedly a challenge, particularly for political reasons, but it is hardly a new one for the alliance.[26] And it is not an especially daunting budgetary challenge.

Consider how these concerns could play out in the transatlantic relationship. It is true that if Europe insists on developing its own RMA force more or less from scratch—using its limited weapons acquisition funds to develop and purchase satellite constellations, stealthy air fleets, and so on—it will probably give short shrift to more rudimentary and inexpensive tech-

22. Lawrence Freedman, *The Revolution in Strategic Affairs*, Adelphi Paper 318, International Institute for Strategic Studies (Oxford University Press, 1998), pp. 70–75.

23. Greg Sherman, "U.S. Warns of Risk to Defence Ties," *The Australian*, July 31, 1998, p. 15.

24. Bryan Bender, "US Worried by Coalition 'Technology Gap,'" *Jane's Defence Weekly*, July 29, 1998, p. 8.

25. William Drozdiak, "Allies Need Upgrade, General Says," *Washington Post*, June 20, 1999, p. A20.

26. Paul B. Stares, *Command Performance: The Neglected Dimension of European Security* (Brookings, 1991), pp. 194–205.

nologies needed to ensure allied interoperability. If it refuses to consider buying U.S.-designed digital radios and computer networks, insisting instead on designing its own, it will make it difficult to operate on the battlefield in an integrated fashion with U.S. forces. Correspondingly, if the United States insists that virtually all RMA-related systems be American-made, it will surely breed resentment that it is trying to monopolize the market in next-generation military technology.[27]

Close collaboration between the allies is needed in the information era. For example, with regard to radios, it is not only a matter of agreeing to common frequencies, but also of using common software that allows radios to encrypt, frequency hop, and share bandwidth in synchronicity. Similar challenges have prevented NATO from establishing common standards in computer hardware and software to date.[28] European countries may need to acknowledge U.S. leadership in information technology, and advanced munitions in some cases, and be willing to buy U.S. systems. But the United States may sometimes need to buy European, Japanese, and Korean technology—or at least enter into relatively equitable collaborative ventures. Chances are that the first network-centric information systems will be designed and built in the United States, given the Pentagon's four-to-one edge over Europe in defense research and engineering spending as well as the strength of the U.S. computer industry. Still, the Pentagon should try to find ways to purchase upgrades or certain follow-on systems from the allies or to team up increasingly with allied firms in joint ventures to do so.

If such a spirit of allied collaboration can be established, money should not be a major constraint on developing an integrated and interoperable high-technology alliance. Once they are developed in research labs and test ranges, the prices of purchasing technologies key to creating a system of systems should generally be modest. Innovations in electronics at the turn of the century are characterized as much by declining price as by increasing

27. David C. Gompert, Richard L. Kugler, and Martin C. Libicki, *Mind the Gap: Promoting a Transatlantic Revolution in Military Affairs* (National Defense University, 1999), pp. 84–85. Consider these concrete examples of the problem: the United States has pulled out of the 155-millimeter autonomous precision-guided munitions program, the modular standoff weapon, terminal guidance warheads for the multiple-launch rocket system, and the NATO identification system; its support for the medium extended air defense system has been tepid and uncertain. See Paul L. Hartman and Craig M. Brandt, "International Armaments Cooperation: Can It Fulfill Its Promise?" *Air Force Journal of Logistics*, vol. 22 (Fall 1998), pp. 12–14.

28. Thomas Cooke, "NATO CJTF Doctrine: The Naked Emperor," *Parameters*, vol. 28 (Winter 1998–99), p. 131.

capacity and power. Improvements to munitions, avionics, sensors, computers, and communications systems can be made relatively inexpensively.[29]

Money will be needed for such an investment strategy, but it can be found without increasing European, Canadian, and East Asian defense budgets. There is no reason for most allied military forces to be anywhere near their present size. If Europe and Japan, for example, need to cut force structure to afford modernization, they can certainly do so. Indeed, with the addition of Poland, the Czech Republic, and Hungary to NATO, the European arm of the alliance now has roughly twice as many individuals under arms as the United States. Were more to emulate the example of Britain, cutting force structure to free up resources for mobility and modernization, many new technology expenses could be afforded.[30]

If it were so hard and expensive to keep up with trends in high technology, the U.S. Marine Corps would surely be shut out of the action. Although it has direct access to various U.S. military satellites, strategic transport, and other critical enabling technologies that the other services or joint military programs purchase, its own funds are quite limited. At $10 billion a year, with only about $1 billion set aside for procurement, it has a budget only one-fourth to one-third the size of several large U.S. allies.[31] Yet the Marine Corps is a major player in military innovation—just as the allied forces can be.

To see just how affordable the so-called RMA should be for major U.S. allies, consider the following shopping basket of technologies:

—advanced radios, computers, and identification-friend-or-foe (IFF) devices for all major vehicles and aircraft in three modern ground divisions;

—advanced radios, avionics, data links, bomb racks and designators for precision-guided munitions, and helmet-mounted displays in three air wings;

—a dozen ground stations, some fixed and some mobile, to integrate and disseminate data and commands, with electronics and computers similar to those of the joint surveillance and target attack radar system (JSTARS) aircraft;

29. Minor procurement also includes trucks, satellites, radars, and a host of other systems, some of them "RMA"-related and some not. See Lane Pierrot and others, "The Costs of the Administration's Plan for the Air Force through the Year 2010," memorandum, Congressional Budget Office, November 1994, pp. 23–24; and Frances Lussier, "The Costs of the Administration's Plan for the Army through the Year 2010," memorandum, Congressional Budget Office, November 1994, p. 11.

30. International Institute for Strategic Studies, *The Military Balance 1997/98*, pp. 293–94.

31. U.S. Marine Corps, *Concepts and Issues 97* (Department of Defense, 1997), pp. 20–25, 134; International Institute for Strategic Studies, *The Military Balance 1997/98*, pp. 293–94.

—a fleet of fifty UAVs of various sizes, ranges, and payloads, similar to those operated by the United States;

—1,000 cruise missiles;

—5,000 short-range munitions, including an assortment of laser-guided bombs, Maverick, and Hellfire-like ordnance;

—500 advanced air-to-air missiles, a mix of beyond-visual-range radar-homing missiles and short-range infrared missiles;

—a squadron of stealth aircraft to map out an enemy's radar defenses and lead any attack, particularly in its first days; and

—several batteries of theater missile defense radars and missiles.

Although admittedly not a comprehensive list, it is nevertheless an extensive one. This basket of capabilities includes most of the types of advanced sensors, computing and communications grids, and precision firepower needed to rapidly detect, target, and destroy military assets on the future battlefield and could apply to a major modernization effort by one of the larger allies of the United States.

In rough numbers, the respective costs of this system of systems would be perhaps $3 billion for the army radios and small computers; $1 billion for the ground stations; $1 billion for the UAVs; $2 billion for the cruise missiles; $500 million for the smaller air-to-ground munitions; $500 million for the advanced air-to-air missiles; $2 billion for the computers, avionics, radios, data links, and helmet-mounted displays for aircraft; $2.5 billion to acquire the stealth aircraft; and $2 billion to $4 billion for missile defenses, for a total investment of around $15 billion.[32] That is hardly an inexpensive price tag. Yet it is the equivalent of less than two years of procurement spending for a major European country or Japan. Averaged out over a decade, it would equal perhaps 15 to 20 percent of the weapons acquisi-

32. For documentation of these costs, see, on theater missile defenses, David E. Mosher, "The Grand Plans," *IEEE Spectrum*, vol. 34 (September 1997), pp. 28–39; on munitions and JSTARS, Michael O'Hanlon, *How to Be a Cheap Hawk: The 1999 and 2000 Defense Budgets* (Brookings, 1998), pp. 132–34; on the costs of partially stealthy aircraft, Donald Stevens and others, *The Next-Generation Attack Fighter* (Santa Monica, Calif.: RAND, 1997), p. 70; on munitions, Michael Miller and others, "Costs of Operation Desert Shield," staff memorandum, Congressional Budget Office, January 1991, p. 17; on cruise missiles, David Mosher and Michael O'Hanlon, *The START Treaty and Beyond* (Congressional Budget Office, October 1991), p. 139; on communications and computers, Mark Hanna, "Task Force XXI: The Army's Digital Experiment," *Strategic Forum*, no. 119 (July 1997), pp. 1–4, and David Pugliese, "Software, Computers to Lead Canadian CF-18 Upgrades," *Defense News*, January 11, 1999, p. 8; on IFF technologies, Brooks Tigner, "NATO Likely Will Adopt U.S. Tank ID System," *Defense News*, January 5, 1998, p. 3; and on UAVs, Defense Airborne Reconnaissance Office, *UAV Annual Report, FY 1996* (Department of Defense, November 1996), and "USAF to Upgrade F-16C/D Avionics," *Aviation Week and Space Technology*, July 6, 1998, p. 59.

tion budget of such a country. Generally speaking, there would be few—if any—additional operating costs, since these systems would replace existing assets rather than require the formation of new units. Moreover, some of these capabilities are already in the hands of European militaries; for example, about fifteen European countries have UAVs of some type today, and many intend to upgrade their capabilities under existing plans.[33]

Even adding in the costs of improving their power-projection capacities in areas such as long-range transport and mobile logistics, the total defense restructuring price tag would not exceed 30 percent of the typical European country's acquisition budget. Since most European militaries are at least 30 percent larger than need be, that price tag is hardly beyond the realm of the possible.[34]

Admittedly, this list of technologies does not push all types of possible military innovation equally rapidly. It focuses on information warfare, not stealth or speed. Pursuing stealth across the board would be extremely costly. Not only combat aircraft but also transport helicopters, ships, and armored vehicles might have to be designed and built from scratch to fully profit from abilities to reduce radar, acoustic, and other signatures. However, these types of improvements are not at the core of the current electronics-led defense modernization wave. They are generally areas where technological progress, albeit impressive, is clearly modest and evolutionary in pace or where radical improvements in capability are unlikely to be necessary.

What about multinational coalitions conducting humanitarian relief or peace operations? Usually, these types of coalitions include militaries with far less preparedness than even the least excellent of the NATO alliance's members. Will future trends in defense technology make it even harder for the United States and close allies to operate with such countries' armed forces?

In the narrow sense, the answer is no. To the extent actual fighting proves necessary, the better militaries from the coalition could simply assume the more challenging tasks, just as they could today. To the extent the South Asians or European neutrals or African states fall even further behind the United States due to new inventions and concepts of operations, that would simply add another reason for them not to participate in the most difficult phases of a mission.[35] Yet they would remain able to patrol, protect fixed

33. Ted Hooton, "European UAVs Take Off," *Armed Forces Journal International,* vol. 136 (July 1999), p. 38.

34. See O'Hanlon, "Transforming NATO."

35. General Anthony Zinni, "It's Not Nice and Neat," *Proceedings* (August 1995), p. 30; Department of the Army, *FM 100-23: Peace Operations* (December 1994); Michael

infrastructure, ensure the safe distribution of food and care to needy populations, and conduct other important jobs even without substantial
enhancements to their weaponry, electronics suites, sensor capabilities, or
other assets frequently associated with the RMA.

This is not to say that countries should be satisfied with their peace operations capabilities just because they prove able to participate in a given mission. It is undesirable that only a modest number of countries can contribute
meaningfully to the most difficult tasks in peace operations, as is the case
today. Among other consequences, that situation means that if the United
States and the top one or two European military powers sit out an operation, as they did in Rwanda in 1994, it will generally be impossible for the
rest of the world to carry it out.

The solution to this problem is not for the world's small and medium
powers to become obsessed with the RMA debate, however, and to become
convinced that only through major purchases of a wide swath of next-
generation technologies and weapons will they be able to contribute to
peace operations. As with allied forces, the greater challenge is probably
their existing shortfalls in power projection and mobile logistics capabilities. Many also need to improve the military preparedness and competence
of troops to conduct basic infantry operations. Some focus on so-called RMA
technologies is appropriate, but a subset of those mentioned previously for
the allies—most notably interoperable radios and computer systems—would
be adequate as a first building block for the world's smaller powers.

Military Innovation and a Nuclear-Free World?

In the last years of the twentieth century, a serious movement proposing the
abolition of nuclear arsenals gained strength and the support of a number
of high-profile proponents. The goal dates to the 1968 Nuclear Nonproliferation Treaty (NPT), premised as it is on the bargain that nonnuclear states
would stay nonnuclear if existing nuclear weapons states eventually elimi-
nated their weapons.

During the cold war and its immediate aftermath, the nuclear superpowers
considered it unrealistic to do much more than try to gradually reduce
nuclear arsenals from their astronomical sizes. Few mainstream security analysts took the exact language of the NPT seriously; after all, a literal read-

O'Hanlon, *Saving Lives with Force: Military Criteria for Humanitarian Intervention*
(Brookings, 1997), pp. 42–46.

ing of that treaty would have required countries to move to "general and complete disarmament," presumably meaning the complete elimination of their conventional military forces, and, arguably, even their internal security forces as well. During the early to mid-1990s, even most nuclear abolitionists focused on more practical goals such as the comprehensive test ban treaty.[36] Another agenda item was introduced as well, the de-alerting of nuclear forces to reduce the risks of accidental launch.[37]

In the very last years of the century, however, what had heretofore been an action item for only the extreme Left became almost a mainstream cause.[38] Among others, former commander of the Strategic Air Command General Lee Butler joined the ranks of the nuclear abolitionist movement.[39] Today there is no shortage of committed proponents.

The motivation of the nuclear abolition movement is clear. Its members worry, not unreasonably, that, if retained, nuclear weapons will someday be used, with catastrophic results for those immediately concerned and perhaps broader swaths of humanity as well. Depending on whether a warhead measured in the range of ten to twenty kilotons (like the Hiroshima and Nagasaki bombs) or more, and depending on where it was dropped, it could kill anywhere from tens of thousands to more than 1 million people. Once one was employed, moreover, there is good reason to fear that more might be used. In that event, economic and societal breakdown could result regionally, leading to widespread famine and disease. If nuclear weapons were used against nuclear power plants, thousands of kilometers of cities, farmland, and water supplies could be badly contaminated by fallout for

36. Coit D. Blacker and Gloria Duffy, eds., *International Arms Control: Issues and Agreements*, 2d ed. (Stanford University Press, 1984), pp. 157, 170; Leonard Weiss, "The Nuclear Nonproliferation Treaty: Strengths and Gaps," in Henry Sokolski, ed., *Fighting Proliferation: New Concerns for the Nineties* (Government Printing Office, 1996), pp. 46–47.

37. See Bruce G. Blair, *Global Zero Alert for Nuclear Forces* (Brookings, 1996); Stansfield Turner, *Caging the Nuclear Genie: An American Challenge to Global Security* (Boulder, Colo.: Westview Press, 1997), pp. 97–106; and Michael J. Mazarr, "Virtual Nuclear Arsenals," *Survival*, vol. 37 (Autumn 1995), p. 8.

38. For recent writings on the topic, see Barry M. Blechman and Cathleen S. Fisher, "Phase Out the Bomb," *Foreign Policy*, no. 97 (Winter 1994–95), pp. 79–95; Jonathan Schell, *The Gift of Time: The Case for Abolishing Nuclear Weapons Now* (Henry Holt, 1998); Robert S. McNamara, *In Retrospect* (Vintage Books, 1996), p. 345; General Andrew J. Goodpaster and General Lee Butler, "Joint Statement on Reduction of Nuclear Weapons: Declining Utility, Continuing Risks," December 4, 1996 (www.stimson.org/zeronuke/generals/j-state.htm [September 1999]).

39. For an important example of a group that began to take the idea very seriously, see Committee on International Security and Arms Control, *The Future of U.S. Nuclear Weapons Policy* (Washington, D.C.: National Academy of Sciences, 1997), pp. 85–98.

years, possibly leading to many more deaths—or, at a minimum, great costs in relocating people and industry.[40]

The political challenges to reaching a nuclear-free world are clearly daunting. But is the idea of a nuclear-free world even plausible on technical grounds? In other words, will there be any way to verifiably eliminate nuclear weaponry worldwide and ensure that new arsenals are not developed without being detected? Moreover, will it be possible to contain the threat of biological weapons sufficiently that nuclear deterrence will not be needed against that potentially horrendous threat?

Within the time frame of this study at least, 2020 or thereabouts, the answers to these questions seem clearly to be no. To begin with, not only are current sensors unable to scan large amounts of territory for nuclear weapons, but nothing on the technical horizon will be able to do so.[41] Natural uranium is far too prevalent on earth for all deposits to be put under reliable controls. With a vast broadening of the International Atomic Energy Agency's efforts, it might be possible to monitor the use of all traditional nuclear reactors and gaseous diffusion plants worldwide.[42] However, a number of other ways of enriching uranium—ranging from centrifuges to calutrons to laser-enrichment devices—are easier to hide. They do require sophisticated equipment, but not necessarily in very large amounts if the goal is to build only a modest number of nuclear devices. [43] Monitoring all relevant technologies would be a daunting and probably impractical task. Proliferating countries might someday be able to produce enrichment or separation technologies indigenously as well, making the monitoring job virtually impossible.[44]

The problem is compounded further by the existing—and, in many cases, unmeasured and unknown—stocks of enriched uranium and plutonium

40. See, for example, Barbara G. Levi and others, "Civilian Casualties from Counterforce Attacks," in Frank von Hippel, ed., *Citizen Scientist* (Simon and Schuster, 1991), pp. 137–49; Bennett Ramberg, *Nuclear Power Plants as Weapons for the Enemy: An Unrecognized Military Peril* (University of California Press, 1980), pp. 68–69.

41. See also Committee on International Security and Arms Control, *The Future of U.S. Nuclear Weapons Policy*, p. 90.

42. Weiss, "The Nuclear Nonproliferation Treaty," p. 42. See also David H. Albright, Barbara G. Levi, and Frank von Hippel, "Fissile Weapons Materials," in von Hippel, *Citizen Scientist*, pp. 105–16. Albright, Levi, and von Hippel persuasively attempt to demonstrate that late twentieth-century technology could detect a uranium enrichment or plutonium extraction plant capable of separating enough fissile material to produce several hundred weapons a year. However, their argument implicitly acknowledges that it could be hard to detect much smaller facilities.

43. Blacker and Duffy, *International Arms Control*, p. 162.

44. For a more optimistic view, see Wolfgang H. Reinicke, *Global Public Policy: Governing without Government?* (Brookings, 1998), pp. 173–217.

around the world (particularly in Russia).[45] Once brought into a verification regime, these stocks could probably be adequately monitored through various types of tags, seals, and old-fashioned security measures—at a minimum providing a warning should they be removed from controls.[46] This holds out promise only for those materials within the regime, however. Such measures may well be worthwhile but should not necessarily be seen as a meaningful step toward complete nuclear abolition.

The basic problem is that, as documented in chapter 3, no sensor in existence or on the drawing board can see nuclear materials beyond a range of a few hundred meters. At greater distances, natural background radiation and other radiation sources would dominate any received signal. So this is not a limit that could be easily extended by an improved sensor or data processing system. Improved computers could extract a fainter signal from a sea of noise with time, but this type of progress would not radically alter detection ranges. Short of a fundamentally different, and at present unforeseen, physical process to search for fissile material or warheads, the basic situation is very unlikely to change appreciably.[47] Moreover, a shielded warhead could be virtually invisible outside the room in which it is stored—a room that may be located ten stories underground.

If sheltered or stored underground, a country's fissile materials or nuclear weapons could be kept hidden even if other countries had unencumbered access to its airspace. Indeed, they might be kept hidden even if the international community had access to the proliferator's city streets and buildings, unless a whistle-blower from the country in question chose to reveal intelligence on the whereabouts of an illicit weapons program or fissile material cache. Whistle-blowers often do emerge from repressive and ruthless regimes, but there is no guarantee one will do so quickly enough to foil his or her government's aggressive designs in a given instance.

The technical argument that it will not become possible to reliably monitor all nuclear materials in the next few decades is a major strike against the nuclear abolitionist cause. There is another strong technical argument. It concerns biological weapons and, as discussed in chapter 4, the enormous

45. Office of Technology Assessment, *Dismantling the Bomb and Managing the Nuclear Materials* (Washington, D.C., 1993), p. 163; Jason Ellis, "Nunn-Lugar's Mid-Life Crisis," *Survival*, vol. 39 (Spring 1997), p. 89.

46. Joseph C. Anselmo, "Defector Details Plan to Plant Nukes in U.S.," *Aviation Week and Space Technology*, August 17, 1998, p. 52.

47. S. T. Belyaev and others, "The Use of Helicopter-Borne Neutron Detectors to Detect Nuclear Warheads in the USSR-U.S. Black Sea Experiment," in Frank von Hippel and Roald Z. Sagdeev, eds., *Reversing the Arms Race: How to Achieve and Verify Deep Reductions in the Nuclear Arsenals* (New York: Gordon and Breach, 1990), pp. 399–402.

threat they could pose in the twenty-first century. Although some would argue from principle that nuclear weapons should not be used to deter biological attack, this argument may prove increasingly difficult to sustain in a century when far more dangerous biological warfare agents may be engineered. As the twentieth century ends, nuclear weapons retain their distinction as the worst weapon of mass destruction, but that distinction may not last more than a couple decades. Far more virulent agents may be developed in the coming years. And the efficient delivery of biological agents, although difficult, is hardly impossible.[48]

Might it be possible to prevent biological weapons proliferation through a combination of improved reconnaissance and arms control, thereby making nuclear deterrence against the biological threat unnecessary? Although reconnaissance and arms control can help, the answer to this question also appears to be no. There is some chance a competent proliferator would be caught—but also a very good chance it would not.

First, finding stocks of biological weapons is extremely difficult. Sensors to find them are, and will remain, restricted to short ranges, as discussed in chapter 3. Indeed, they generally will be extremely short range, requiring immediate physical access to the containers holding the weapons, unless the biological agents are released into the air or otherwise dissipated. This is because the lasers and other electromagnetic beams that could be directed at suspicious agents—thereby producing patterns of absorption and emission from the aerosols that might convey information about their identity—cannot penetrate most buildings, vehicles, and containers where the agents would normally be stored.[49]

In that regard, biological weapons pose a challenge similar to that of nuclear weapons. In another regard, they are even more problematic, for once initial specimens are available, they can be reproduced quickly and in very small facilities using a minimum of sophisticated equipment.

There might be ways to distinguish between laboratories conducting allowed defensive research and those conducting production for weaponization, based on the amount of agent and how it is treated and tested. Limits on allowed defensive research could also be tightened. But making such a

48. For a different argument—though one that concedes that nuclear weapons may have a certain "existential deterrent" quality against biological weapons—see Harold A. Feiveson, ed., *The Nuclear Turning Point* (Brookings, 1999), pp. 335–41.

49. For example, see the abstracts and papers from the Fourth Joint Workshop on Standoff Detection for Chemical and Biological Defense, organized by the Department of Defense's Joint Science and Technology Panel on Chemical and Biological Defense at Hospitality House, Williamsburg, Va., October 26–30, 1998.

system work would be very hard. Finding relevant facilities in a timely fashion so as to be able to inspect them is difficult, as the United Nations Special Commission (UNSCOM) on-site experience in Iraq has demonstrated.[50] Moreover, it may be possible for a proliferator to destroy evidence of biological weapons production within a few hours' time—quite possibly before inspectors could hope to arrive.[51]

On-site inspection regimes are good enough to catch some violators, deter others, and limit the scope of cheating by yet others. For these reasons they are generally desirable. Yet their limitations must not be forgotten, particularly when extreme proposals, such as the complete abolition of nuclear and biological agents, are being considered.[52] Nor will foreseeable detection technologies, all of which require immediate access to a biological agent unless that agent has been released into the air or atmosphere, make any fundamental difference.

Some would argue that U.S. military superiority is great enough to permit a strictly conventional military response to any battlefield use of weapons of mass destruction against American forces.[53] Whatever degree of U.S. military superiority exists today, however, there is no guarantee it will last forever. Even today, there are important uncertainties about just how well the U.S. military could cope with a full-fledged chemical or biological attack.[54] In addition, fixed bases upon which American forces depend heavily, such as airfields, could clearly be heavily damaged in a nuclear attack. The U.S. military advantage over most potential foes is great enough that American forces could probably prevail without resorting to the use of nuclear weapons. However, they would most likely do so at the price of high U.S.

50. See Milton Leitenberg, "Biological Weapons Arms Control," PRAC Paper 16 (University of Maryland at College Park, Center for International and Security Studies, 1996), pp. 23–39, 69–84.

51. Perhaps 3,000 sites in the United States alone would need to be declared under a biological weapons inspection regime. See Jonathan B. Tucker, "Verification Provisions of the Chemical Weapons Convention and Their Relevance to the Biological Weapons Convention," in Marie I. Chevrier and others, *Biological Weapons Proliferation: Reasons for Concern, Courses of Action*, Report 24 (Washington, D.C.: Henry L. Stimson Center, January 1998), pp. 77, 93–97.

52. Amy Smithson, "Man versus Microbe: The Negotiations to Strengthen the Biological Weapons Convention," in Chevrier and others, *Biological Weapons Proliferation*, p. 111.

53. See, for example, Steve Fetter, "Limiting the Role of Nuclear Weapons," in Feiveson, *The Nuclear Turning Point*, p. 40.

54. Victor A. Utgoff, *The Challenge of Chemical Weapons: An American Perspective* (St. Martin's Press, 1991), pp. 148–88; Terry J. Gander, ed., *Jane's NBC Protection Equipment* (Alexandria, Va.: Jane's Information Group, 1995); Cohen, *Report of the Quadrennial Defense Review*, pp. 24, 49.

casualties. Rather than accept high casualties, the United States would have powerful reasons to use nuclear weapons against an enemy's forces and military infrastructure in response, both to save its own troops' lives and to deter further enemy attacks of that kind in the future. Making the possibility of such a response known in advance, as it did before Operation Desert Storm, could also have deterrent benefits. It could discourage a foe from the belief that by threatening to use weapons of mass destruction against U.S. forces it could keep the casualty-averse United States from responding to its aggression.[55]

Conclusion

The French frequently accuse Americans of believing that technology supercedes politics in human affairs and that many problems can be fixed with the right gadget or invention. The French may overstate their case; after all, American high technology did win the 1991 Gulf War in a resounding fashion and did much to ensure the Western world's success in the cold war as well.

Yet for the policy questions considered in this chapter, the French criticism of U.S. tendencies is correct. Those who subscribe to the notion of a revolution in military affairs often expect that it will transform not only defense technology, tactics, and operations, but also broader matters of national strategy. For example, many are wont to claim that overseas military bases will become far less important in the future. They also frequently assert that it will become more difficult for allied forces to be tightly integrated and interoperable in the future. Finally, some claim that the advent of better verification technologies, as well as increasingly lethal and accurate conventional military munitions, will make it possible to eliminate nuclear weapons from the face of the earth within the next few decades.

On all these points, the RMA proponents who advance them are wrong, particularly for the time horizon of ten to twenty years that is of direct relevance to defense planning. No technologies on the drawing boards, laboratories, or test ranges will permit the United States to become an isolationist global power, basing the vast preponderance of its military on its own territory and striking intercontinentally when necessary. No technologies are within reach to allow nuclear materials to be reliably detected around the world. More positively, nothing about current innovation in defense tech-

55. See Mosher and O'Hanlon, *The START Treaty and Beyond*, pp. 15–24.

nology and tactics will make it impossible for the allies of the United States to work together effectively with the American military in the future. If it becomes possible to do without nuclear weapons or overseas bases, and if allied forces wind up being incapable of cooperating with those of the United States, these will be primarily the successes and failures of politics—not the product of a revolution in military affairs.

CHAPTER SEVEN

A Defense Modernization Strategy

In their belief that we have reached a turning point in military technology, tactics, and strategy, many revolution in military affairs (RMA) proponents would have the United States spend historically disproportionate amounts of the Pentagon budget on hardware to transform the U.S. armed forces. Ironically, this recommendation comes at a time when U.S. dominance in military technology is more overwhelming, and less challenged, than at any moment in the nation's history. The rogue states of Iraq, North Korea, and Iran possess largely obsolescent military equipment and together spend less than one-tenth as much as the United States does each year modernizing their armed forces (see figure 7-1). Russia's economy is structurally unsound and stagnant.[1] Most of its industries produce outputs worth less than the sum of their inputs. China is still growing fast, but its national banking problems are enormous, verging on insolvency. At the same time, greatly increased amounts of resources are needed to improve the country's infrastructure and agriculture.[2] China's military readiness is also mediocre, and its weapons industries resemble fiefdoms more than an integrated defense industrial base.[3] Not only are these

---

1. Clifford G. Gaddy and Barry W. Ickes, "Russia's Virtual Economy," *Foreign Affairs*, vol. 77 (September–October 1998), pp. 53–67.

2. Nicholas R. Lardy, *China's Unfinished Economic Revolution* (Brookings, 1998), pp. 15–20, 183–222.

3. Richard A. Bitzinger and Bates Gill, *Gearing up for High-Tech Warfare? Chinese and Taiwanese Defense Modernization and Implications for Military Confrontation across the Taiwan Strait, 1995–2005* (Washington, D.C.: Center for Strategic and Budgetary Assessments, 1996), pp. 16–24; Bates Gill and Michael O'Hanlon, "China's Hollow Military," *National Interest*, no. 56 (Summer 1999), pp. 55–62.

Figure 7-1. *International Distribution of Military Spending,*
Selected Countries, 1998

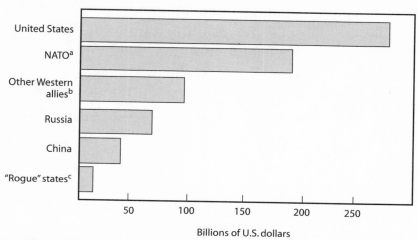

Billions of U.S. dollars

Source: International Institute for Strategic Studies, *The Military Balance 1998/99* (Oxford University Press, 1998), pp. 295–99; and Michael O'Hanlon, *How to Be a Cheap Hawk: The 1999 and 2000 Defense Budgets* (Brookings, 1998), p. 29.

a. Includes Poland, Hungary, and the Czech Republic, but not the United States.

b. Includes Japan, South Korea, Australia, New Zealand, Thailand, Phillipines, and the Rio Pact countries minus Cuba (including all South American countries except Belize, Guyana, and Suriname, and also four Caribbean islands or island groups: the Bahamas, the Dominican Republic, Haiti, and Trinidad and Tobago).

c. Includes Cuba, North Korea, Iran, Iraq, and Libya.

countries unable to rival U.S. military power today, they will be unable to do so for decades—even in battles fought in their own regions of the world.

While some RMA proponents perform a valuable service by challenging the Pentagon's existing weapons modernization agenda, most share the Department of Defense's (DoD's) view that investment in technology must be radically increased across a wide range of weapons and technologies. This poses a major fiscal problem. According to the Congressional Budget Office, DoD's current modernization plans will require annual real procurement budgets of at least $70 billion and perhaps as much as $90 billion (as expressed in year 2000 constant dollars)—in contrast to the annual average of around $50 billion that characterized the late 1990s.[4] Many RMA proponents, in their affinity for technology, offer proposals for more hard-

4. See Lane Pierrot, statement on Aging Military Equipment before the Subcommittee on Military Procurement of the House Committee on Armed Services, 106 Cong. 1 sess., February 24, 1999; Office of the Under Secretary of Defense (Comptroller), *National Defense Budget Estimates for FY 2000* (Department of Defense, 1999), pp. 124–29.

Figure 7-2. *U.S. Defense Procurement Spending, 1945–2003*

Percent of total Pentagon outlays

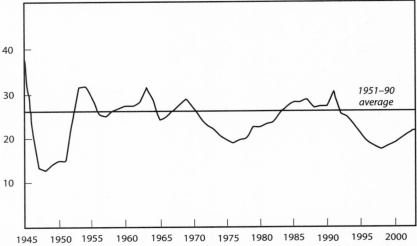

Source: Office of the Under Secretary of Defense (Comptroller), *National Defense Budget Estimates for FY99* (Department of Defense, March 1998), pp. 112–17.

ware that could further exacerbate this problem. In light of its "procure-ment holiday" in the 1990s, when it purchased relatively little weaponry (see figure 7-2), the Pentagon does need to spend more money buying equip-ment. Many old systems are wearing out and require replacement just to keep equipment safe and reliable. But the Pentagon needs to find ways to prioritize, and keep its appetite for new weaponry in check, given the unlike-lihood that defense budgets will increase enough to fund its entire mod-ernization agenda. If the RMA movement does not help the Pentagon to prioritize, it will have made the fiscal problem worse rather than better.

If budget surpluses in the United States permit real defense spending to increase from its 1999 level of about $280 billion to more than $300 bil-lion (in constant 2000 dollars), the Pentagon may be able to have it all (see figure 7-3). The odds that real defense spending will increase by that amount seem low, however. Although defense spending may approach $290 billion in 2000, Congress had to use generous definitions of "emergency spending" to fit so much expenditure under the caps of the balanced-budget agreement and may not be able to repeat the budgetary gamesmanship to the same degree in the future. Revealingly, the purportedly pro-defense Congress based its 1999 tax cut proposal on a budget plan that would have had real defense spending *decline* every year after 2004, winding up with lower

Figure 7-3. *U.S. National Defense Spending, 1962–2004*

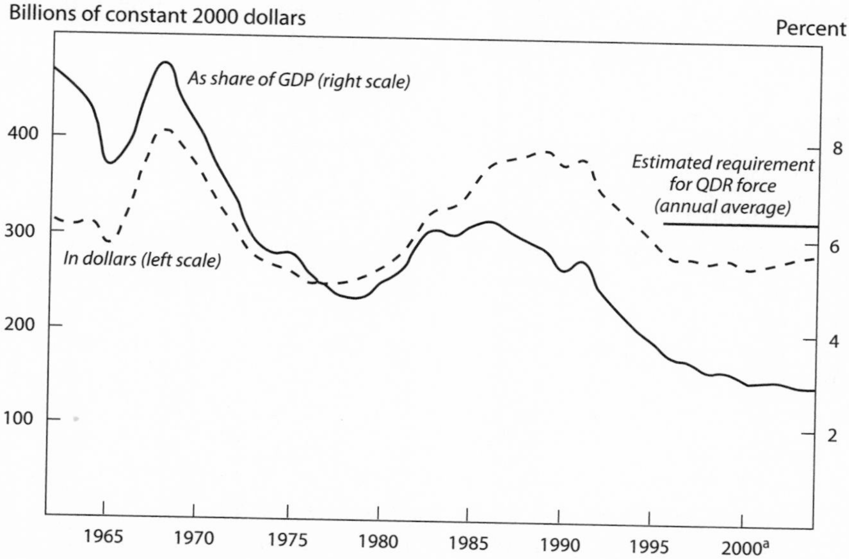

Billions of constant 2000 dollars

Source: Executive Office of the President, *Budget of the United States Government: Historical Tables, Fiscal Year 2000* (February 1999), pp. 118, 120.

a. Figures from 2000 onward are the president's estimates based on the FY00 budget proposal.

b. Annual average. Author's estimate of spending required to sustain the QDR force.

spending levels than in 1999. In that event, the Quadrennial Defense Review (QDR) plan would be simply unaffordable. Even under Clinton-Gore long-term defense spending assumptions, a less expensive alternative to the QDR would probably be needed.

Some RMA proponents recognize that defense spending is unlikely to increase substantially, but still think that substantial additional resources must be found to reequip and restructure the U.S. armed forces. They often propose taking the necessary money out of near-term military readiness, perhaps also scaling back the size of the active-duty force by a large percentage. This would be unwise, however. In today's world, the United States should sustain a defense strategy emphasizing forward presence operations, exercises with allies and neutrals, military-to-military exchanges, peacekeeping missions, and response to crises when necessary. These are the tools with which alliances are kept credible and cohesive, and potential adversaries are contained or deterred. They are the linchpins in an alliance system accounting for more than 75 percent of global gross domestic product (GDP) and global military spending. Ensuring that the Western community of industrialized nations remains cohesive and credible is more important than rushing to further widen the already huge gap between American military capabilities and those of any other country.

Nonmilitary tools of foreign policy, such as foreign aid, also have an important role in this broader security policy and are being underfunded today.[5] Raising their very modest budgets to slightly greater levels would be further complicated by a Pentagon encouraged by RMA proponents to lay its hands on a large share of the budget surplus.

This chapter examines how to promote defense innovation at current Pentagon budget levels without harming U.S. military engagement and deterrence around the world. It draws on the technical analyses of previous chapters to fashion a defense modernization plan that is less costly than the Pentagon's own plan but that promotes key technologies and critical research and experimentation efforts. It is built around four guidelines:

—Emphasize relatively economical and high-payoff improvements in munitions, communications, information systems, and sensors that are possible today due to trends in electronics and computers.

—Redress existing military weaknesses and vulnerabilities. Doing so requires attention to a wide spectrum of subjects ranging from homeland defenses against missile attack and the terrorist threat to further improvements in U.S. airlift and sealift.

—Sustain robust research, development, testing, and evaluation (RDT&E) efforts, particularly in areas of basic research and development (R&D), as well as joint-service simulation and experimentation.

—Do not pursue full-scale modernization with expensive next-generation weapons platforms. Rather, focus on making sure platforms remain safe and reliable. In many cases, weaponry that is already in the force today will be good enough for coming decades (though it may have to be refurbished, or new production runs may have to be authorized, to keep equipment young enough to be dependable).

The first three pillars of this modernization strategy will take more money than the Pentagon now plans. The fourth pillar is the bill payer, making possible the first three elements of the proposed modernization agenda at current real spending levels.

This modernization agenda still would not be cheap. It would require at least $65 billion a year in real procurement spending into the indefinite future,

5. See Mickey Edwards and Stephen J. Solarz, *Financing America's Leadership: Protecting American Interests and Promoting American Values* (New York: Council on Foreign Relations, 1997); Michael O'Hanlon and Carol Graham, *A Half Penny on the Federal Dollar: The Future of Development Aid* (Brookings, 1997).

as well as a sustained RDT&E budget of more than $30 billion (these figures are expressed in terms of constant 2000 dollars). However, when coupled with changes in the two-war strategy, nuclear force posture, and naval forward presence operations that I have suggested elsewhere, it would allow about $20 billion a year in savings relative to the QDR plan.[6] That would allow it to be sustained at year 2000 real defense spending levels, rather than be dependent on a large peacetime increase in real defense spending that neither political party seems likely to deliver.

Emphasize Electronics and Computers

The U.S. military certainly should modernize its weaponry, even in this era of overwhelming American defense dominance. But it need not do so precipitously or expensively. By focusing on military capabilities that make maximum use of modern electronics and computers, combat capabilities can be improved impressively at modest cost. This philosophy could be termed the systems of systems approach to military modernization, as it places emphasis less on major weapons platforms than on what they carry and how they are networked. One can categorize the most important technologies into two groups: advanced precision munitions on the one hand and information networks involving sensors and communications systems on the other.

There are several priorities in the area of munitions. One is to keep emphasizing R&D of new capabilities with various combinations of sensors, such as infrared homing and acoustic devices. We do not yet know how new forms of homing submunitions will perform in the face of decoys, complex terrain, and other battlefield realities. As such, it is important to look at a number of types, and to conduct vigorous testing and experimentation. Thankfully, the Pentagon seems to agree with this objective, as reflected in its programs for the sensor-fuzed weapon (SFW), brilliant anti-tank (BAT) munitions, and other advanced ordnance.

However, the Pentagon's commitment to fund adequate amounts of precision munitions is less solid. David Ochmanek and coauthors make a convincing case for purchasing some 15,000 munitions for a two-war capability plus war reserve stocks. That estimate accords with the fact that the forces of countries like Iraq and North Korea typically possess up to 10,000 armored vehicles each. Not all would have to be destroyed in a war, but

6. See Michael O'Hanlon, *How to Be a Cheap Hawk: The 1999 and 2000 Defense Budgets* (Brookings, 1998), pp. 36–41.

perhaps 50 percent would be in order to confidently ensure enemy defeat; weapons' kill probabilities might be anywhere from 0.25 to two or three (kill probabilities greater than one correspond to munitions carrying multiple submunitions). This numerical goal is more than double what the services now plan to acquire in the form of weapons such as the SFW and joint standoff weapon (JSOW) carrying skeet submunitions, but it is not that expensive to attain. Assuming an average unit cost of $250,000 per munition (the current production cost of the sensor-fuzed weapon), it translates into an added price tag of some $2.5 billion, or a few hundred million dollars a year.[7]

In the realm of communications, the watchwords should be joint-service interoperability, robustness, redundancy, and hardening. These systems are the linchpins of American military technological excellence and will prove even more critical in the future. Yet electronics are vulnerable to physical disruption, including radio-frequency weapons and high-altitude electromagnetic pulse (EMP) from nuclear bursts as well as old-fashioned jamming. Software is vulnerable to crashes and hackers. And the military services still pay insufficient heed to the requirements of operating together when designing their information technology and communications systems. Everything possible must be done to mitigate these shortcomings and vulnerabilities.[8]

The U.S. military is vigilant in protecting electronics and communications in some areas today. For example, even as it develops digitized divisions, it is not throwing away old time-tested technologies like voice radios and hard-copy maps. It also anticipates the possibility that global positioning system (GPS) signals will be jammed and other advanced sensors spoofed or attacked. However, it seems careless in other areas. For example, the Pentagon has drastically curtailed expenditures on hardening electronics against HEMP. Yet today's advanced electronics are far more vulnerable to HEMP than are the older systems of several plausible U.S. foes. The Pentagon's increasing efforts to purchase (generally unprotected) commercial off-the-shelf technologies only reinforce and exacerbate this trend. Moreover, the types of nuclear bursts needed to generate electronics-killing HEMP take place above the atmosphere and do not kill people; as such, the barriers to their use are far less than those to the employment of weapons of mass destruction against population centers or troop concentrations. That is not

7. David A. Ochmanek and others, *To Find, and Not to Yield* (Santa Monica, Calif.: RAND, 1998), pp. 39–40, 76, 94.

8. For a similar view, see Computer Science and Telecommunications Board, National Research Council, *Realizing the Potential of C4I: Fundamental Challenges* (Washington, D.C.: National Academy Press, 1999), pp. 1–26.

to say that an enemy would be cavalier about using a nuclear weapon to generate HEMP; however, if it relied on older and hence hardier equipment itself, it might be sorely tempted. For this reason, the United States should seriously consider restoring its spending on electronics hardening to cold war levels—reportedly several hundred million dollars a year more than are now spent on the purpose.

Funds will be needed in coming years to purchase internet capabilities for army divisions and combat units of the other services. Moreover, given the pace of change in electronics and computer capabilities, one should plan to upgrade Internet capabilities fairly frequently—perhaps every five to ten years, in contrast to the normal cycle of defense innovation, which is typically closer to twenty years.

How much will these capabilities cost? The funds needed for technology development are relatively modest. For example, the army's tactical Internet, or FBCB2 system, is estimated to cost some $400 million to develop over eight years.[9] To purchase the necessary hardware for a division might cost around $1 billion, meaning that the army would need about $10 billion over ten years to outfit its entire force. Demands in the other services would be of comparable scale. Some are already programmed into service budget plans, but an additional $1 billion to $2 billion a year might be needed DoD-wide.[10]

Added impetus should also be given to joint-service communications and computers. Today the various U.S. military services can talk to each other by voice radio, but they are making only very slow progress toward being able to exchange targeting data and related digital information. An anecdote is illustrative of the problem today: If an advanced fighter jet is conducting close-air support for ground units engaged in combat, it may literally need to be told where to find the target by an army soldier on the ground. The communication is still often by voice and often involves directives like "do you see the third building on the second hill on the other side of the river? If so, please shoot at it." This message would be given to a pilot and

9. General Accounting Office, *Battlefield Automation: Acquisition Issues Facing the Army Battle Command, Brigade and Below Program*, GAO/NSIAD-98-140 (June 1998), pp. 7–8; James R. Blaker, "Revolution(s) in Military Affairs: Why the Critique?" *National Security Studies Quarterly*, vol. 5 (Winter 1999), pp. 86–87; Daniel G. Dupont, "ASD (C3I) Shop Directs Services to Put Independent Radio Efforts on Hold," *Inside the Pentagon*, September 10, 1998, p. 1; Daniel G. Dupont, "DoD Proposes Additional $270 Million for Joint Tactical Radio System," *Inside the Pentagon*, November 12, 1998, p. 1; Daniel G. Dupont, "DSB Says Joint Tactical Radio System Can Be 'Aggressively' Accelerated," *Inside the Pentagon*, March 11, 1999, p. 1.

10. Mark Hanna, "Task Force XXI: The Army's Digital Experiment," *Strategic Forum*, no. 119 (July 1997), pp. 1–4.

crew flying at 5,000 to 20,000 feet of elevation (or higher, depending on the local surface-to-air missile threat). What should happen, and what is already technologically within reach today, is to have the ground controller use a GPS receiver to establish his own position, use a laser or other beacon to determine the coordinates of the target, and then convey those coordinates to the aircrew either by voice or (better yet) by digital link. Appropriately, the assistant secretary of defense for command, control, communications, computers, and intelligence (C4-I) recently seized greater control over the radio programs of the individual services. They must be made more compatible. Such changes depend less on adding money to the Pentagon budget than on spending it more carefully.

DoD's plans to purchase improved reconnaissance capabilities such as enhanced spy satellites and new fleets of unmanned aerial vehicles (UAVs) appear generally sensible and should be sustained. In a few cases, however, added funds may be appropriate. For example, the case for restoring the planned purchase of joint surveillance and target attack radar system (JSTARS) reconnaissance aircraft to nineteen planes from its reduced level of thirteen appears strong. That would add another $200 million or so a year to a ten-year program.

Some procedural changes to promote jointness may be appropriate as well. One sensible idea is that the vice chairman of the joint chiefs of staff should testify directly before Congress, as suggested by the former vice chairman Admiral William Owens. This individual runs the Joint Requirements Oversight Council (JROC), which approves major acquisition programs, and often has a more ecumenical perspective than most uniformed officers.[11] Tom Davis has suggested the composition of the JROC might also be reconsidered to include deputy commanders of the unified and specified commands in addition to the vice service chiefs.[12]

All in all, the Pentagon needs to spend about $3 billion a year more on so-called C4-I technologies and advanced munitions. It also needs to coordinate more of these types of efforts and purchases at the joint-service level.

Redress U.S. Vulnerabilities

Today the United States has several notable military vulnerabilities and weaknesses. Some pertain to how it would deploy forces overseas, some to

11. John Robinson, "JCS Vice Chair Should Testify as Part of Budget Process—Owens," *Defense Daily*, March 1, 1996, p. 312.

12. M. Thomas Davis, "The JROC: Doing What? Going Where?" *National Security Studies Quarterly*, vol. 4 (Summer 1998), pp. 21–42.

defending U.S. territory. Whereas the hardening of key electronics systems in weapons' computers, communications systems, and fire-control mechanisms can address a number of these problems, as argued previously, other steps are needed as well. Total added costs from these various initiatives would reach another $3 billion a year—hardly cheap but much less than what can be saved by streamlining major weapons purchase plans.

One of the troubling dimensions of Desert Shield and Desert Storm, as frequently noted by commentators, was the U.S. need to have several months of buildup time, even in such a well-prepared theater. This was largely due to a lack of adequate pre-positioning in the theater and inadequate surge sealift. The Clinton administration has improved U.S. sealift and overseas pre-positioning of equipment, but more needs to be done, as evidenced by the fact that many U.S. officials believed three months of preparation would have been needed to mount a ground invasion of Kosovo during NATO's 1999 war against Serbia, even though only 40,000 Serbian troops were deployed in Kosovo at the time.[13]

Within about a month, the U.S. military can rapidly deploy one light division, send over troops to marry up with one to two brigades of pre-positioned heavy equipment, and then sealift two additional full heavy divisions to either northeast or southwest Asia.[14] A full five-division corps would be delivered within roughly three months, as fast sealift ships completed a second voyage to the region and slower sealift ships were loaded and deployed.[15]

This is not quite good enough, however. It would be more prudent for the United States to be able to establish a strong holding force in both Korea and the Persian Gulf region quickly. With faster sealift, the United States could deploy two modest-size forces overseas within a month. Each would be roughly patterned after the 200,000-troop Desert Shield operation, which was designed to deter Saddam from invading Saudi Arabia, rather than the 550,000-troop deployment (plus more forces from other countries) later used to expel him from Kuwait. This enhanced transportation capability would also provide insurance against the real possibility that some ships could be

13. Thomas E. Ricks, "Gung-Ho but Slow: Why the U.S. Army Is Ill-Equipped to Move Troops Quickly into Kosovo," *Wall Street Journal*, April 16, 1999, p. A1; John Barry, "Why Troops Take Time," *Newsweek*, April 26, 1999, p. 30.

14. The pre-positioning programs have been expanded considerably and, despite some glitches, are working well. Army brigade sets of material have been added in the Pacific region and in the Middle East, with another due to be in place soon; the reliability of the equipment in these sets is typically at least 90 percent. See General Accounting Office, *Overseas Presence*, GAO/NSIAD-97-133 (June 1997), p. 22; General Accounting Office, *Afloat Prepositioning*, GAO/NSIAD-97-169 (July 1997), pp. 3–5.

15. Rachel Schmidt, *Moving U.S. Forces: Options for Strategic Mobility* (Congressional Budget Office, February 1997), pp. 29, 79.

lost, or forward-stationed supplies destroyed, due to enemy action from mines, submarines, missiles, or commandos.

As one approach, the Pentagon could build more large, medium-speed, roll-on/roll-off (LMSR) vessels for this purpose. However, these ships are expensive, put many U.S. eggs in one basket in places where shallow-water mines and submarines as well as shore-based missiles could sink a ship or two, and could clog up many of the world's ports. Instead, DoD should purchase somewhat slower yet much cheaper roll-on/roll-off ships like those now in the ready reserve. About fifty medium-size transport ships could suffice for two more divisions' worth of combat forces and some initial supplies. At $35 million per ship, that would amount to less than $2 billion, in contrast to the likely cost for twenty LMSRs of about $6 billion.[16]

Airlift should also be increased, with the goal of deploying enough troops, aircraft, and other supplies for two Desert Shield–like operations within a month of the beginning of a crisis. With roughly fifty 747-class aircraft added to its planned inventory, the air force could achieve that goal. It could achieve a 60 million-ton-miles a day (MTM/D) deployment rate rather than the current 50 MTM/D capability. The associated total cost would be about $10 billion.[17] Spread over a decade, the added costs from these sealift and airlift initiatives would be about $1.25 billion annually.

Another $2 billion would be well spent on two types of technologies that are already within reach but that have not been pursued by the U.S. military services as of yet: a lighter tank and a mine-resistant infantry fighting vehicle. The former technology was under development as the army's armored gun system (AGS) until the program was terminated in the mid-1990s due to budget constraints and concerns about the AGS's lethality and survivability in high-intensity combat operations. Its capabilities would nonetheless have been welcomed in a ground war in Kosovo, where it could have allowed U.S. forces to approach potential ambush zones with heavy protection. Likewise, a wheeled mine-resistant vehicle with independent suspensions for each tire and the ability to absorb strong shocks from below would be desirable in the types of settings that are common in places such as Kosovo. Today the army and Marine Corps count too much on breaching minefields during major maneuvers—a tactic more appropriate for a

16. Schmidt, *Moving U.S. Forces*, pp. 26–29.

17. Jeffrey Record, "After the C-17: Coming Airlift Realities and Choices," *Armed Forces Journal International*, vol. 134 (December 1996), pp. 32–33; Schmidt, *Moving U.S. Forces*, p. 17.

mission such as penetrating Iraqi front lines in Kuwait than for operating in small units throughout a country where mines could be anywhere.[18]

Secretary William Cohen's initiatives to improve U.S. naval mine warfare capabilities should be sustained. He has increased the budget for mine warfare from roughly $500 million a year to twice that amount, and it is to remain at the higher level through 2004. This has been a matter of considerable interest to the secretary throughout his tour of service, and rightly so. U.S. war plans continue to hinge, somewhat unrealistically, on the premise that waterways will be safe in the early part of a conflict, making it possible to sail in maritime pre-positioned ships within days and sealift ships within weeks. These assumptions are important to war plans, yet probably not realistic at present—particularly with only two of the U.S. Navy's mine warfare ships normally deployed in Japan and two in the Persian Gulf. Peacetime patrols in confined waters such as the Persian Gulf are also dangerous, increasing the navy's interest in having dedicated mine-hunting capabilities on most ships.

Adding money to mine warfare accounts is no guarantee of success, and indeed appropriate technologies are proving difficult to develop, particularly for shallow-water and buried mines. In the words of marine Major General Dennis Krupp, director of expeditionary warfare, "we're on a fishing trip" in the search for appropriate technologies. Still, there are promising developments in a number of areas. Improvements in the last ten to twenty years include glass-reinforced ship hulls as well as explosive charges for use from landing-craft air cushion (LCAC) amphibious ships in surf zones. Other developments include blue-green lasers and robotic sonars for mine hunting, as well as minesweeping ships that can imitate the magnetic and acoustic signatures of ships more truly, thereby detonating even relatively smart mines.[19]

Finally, any study of future challenges to U.S. security cannot focus exclusively on the battlefield as traditionally defined. To its credit, the Senate Armed Services Committee took a step in this direction in 1999 by creating a new subcommittee on emerging threats and capabilities. Of particular note are possible threats to the American homeland from terrorism and weapons of mass destruction. These types of threats have increased in the

18. See Michael O'Hanlon, "Sins of Omission," *Washington Times*, May 24, 1999, p. A19.
19. Hector J. Donohue, "Minesweeping + Mine Hunting = Success," *Proceedings* (March 1998), pp. 52–55; Buzz Broughton and Jay Burdon, "The (R)evolution of Mine Countermeasures," *Proceedings* (May 1998), pp. 55–58; Roman Schweizer, "Countering Very Shallow Water Threats a Top Priority for Marine Corps," *Inside the Navy*, May 3, 1999.

1990s, most notably in the attack on the World Trade Center towers and the Oklahoma City bombings, not to mention the sarin gas subway attack and earlier attempted biological agent attack against the Imperial Palace by Aum Shinrikyo in Tokyo.

Beginning with missile defense, a subject treated in chapter 4 as well, the increased amounts of money added to theater and national defense programs in the late 1990s seem prudent. The 1997 QDR added $2 billion to the future years' defense plan for national missile defense (NMD), and Congress added another $1 billion for missile defense to the 1999 budget. Although some Clinton administration officials objected to this increase as wasteful, others, including the Pentagon's director of defense research and engineering, conceded that good use could be made of at least some of it. However, it would be a mistake to assume that even greater annual spending levels will speed success. A number of missile defense programs will need more testing than is currently planned. It may not be prudent to accelerate the pace of testing, however; it may be necessary instead to keep testing for a longer period of time, so that mistakes can be diagnosed and corrected as they occur.[20]

The Clinton administration's recent decision to place nearly $7 billion more in the future years' defense program for deployment of an NMD is sound as well. That translates into roughly $1 billion a year, making a grand total of about $1.5 billion a year when added to previously planned amounts for R&D. Deployment costs could easily increase by 50 percent due to normal cost growth in advanced technology systems—or even by more than 100 percent if deployments are made at several sites rather than just one. These costs are considerable but not prohibitive given the stakes involved.[21]

The likely limitations of NMD need to be kept in mind. It is unlikely to be effective against the missile force of any country capable of testing its missiles frequently enough to develop good countermeasures. Although developing good countermeasures is not particularly complicated, especially when compared with making the rockets carrying them and nuclear warheads accompanying them, they may not work properly unless first tested under realistic flight conditions. And a country like North Korea may not have the resources or the international leeway to test its missiles properly. Thus a light nationwide defense might work against such a rogue state—

20. Bryan Bender, "USA to Put Extra $1b into Missile Defence Systems," *Jane's Defence Weekly*, October 28, 1998, p. 2; General Accounting Office, *National Missile Defense*, GAO/NSIAD-98-153 (June 1998), pp. 1–19.

21. Frank Wolfe, "Pentagon to Delay NMD Deployment Decision," *Defense Daily*, December 10, 1998, p. 5.

though it also might not. In addition, an idea of Richard Garwin's for boost-phase missile defense using high-speed ground-based interceptors deserves attention. The downside to such an approach is that the interceptors would have to be based near the country potentially launching the missiles. The major advantage, however, is that these interceptors would destroy outgoing missiles before they could dispense countermeasures.[22]

The United States also needs better capabilities for response to the use of weapons of mass destruction on American soil. Breakthroughs in the same types of biological and chemical sensors that are needed on the battlefield would be useful for cities as well. Funds are also needed to provide better protection equipment and training for National Guard units and special defense squads that can help local police, fire officials, and health crews handle any emergencies. Money is also required for stockpiling antibiotics and mitigating vulnerabilities in large buildings' air circulation systems (such as making sure that the location of vents does not pose an unduly attractive and attainable target for terrorists). More vaccines should be stockpiled as well; for example, at present there are only 7 million doses of smallpox vaccine available in the United States, and producing enough for the entire country could take thirty-six months from the time a decision was made to do so.[23]

In his year 2000 budget request, President Clinton proposed spending more than $1 billion in additional money for some of these initiatives. However, his proposal appears not to include enough funds for certain types of medical preparations, vaccines, and antibiotics. It also fails to address other issues such as the vulnerabilities of prominent private-sector buildings.[24] A detailed agenda for homeland defense is beyond the scope of this book, but a further annual investment of $500 million above the president's recent initiative appears wise.

Sustain Robust RDT&E and Experimentation

My alternative modernization strategy would make no economies in research, development, and experimentation. In fact, it would increase fund-

22. Richard L. Garwin, "Effectiveness of Proposed National Missile Defense against ICBMs from North Korea," March 17, 1999 (www.fas.org/rlg/990317-nmd.htm [September 1999]). See also Michael O'Hanlon, "Rethinking Star Wars," *Foreign Affairs*, vol. 78 (November–December 1999).

23. Tim Beardsley, "Facing an Ill Wind," *Scientific American*, April 1999, p. 20.

24. Office of the Press Secretary, the White House, "Fact Sheet: Funding for Domestic Preparedness and Critical Infrastructure Protection," January 22, 1999.

ing for experimentation—especially for efforts that involve more than one military service—and sustain real funding levels for basic R&D in contrast to Pentagon plans to cut such funding.

Given its current overwhelming technological edge, and the implausibility that any country will challenge it in that regard in the near to medium term, the United States can afford to take an economical approach to procurement as outlined elsewhere in this chapter. However, it has no such luxury with regard to RDT&E. Even if the next U.S. military does not need to be rapidly and radically transformed, we cannot yet say the same for the "military after next." In technological as well as geostrategic terms, the world beyond 2020 is unpredictable, and the United States must be prepared for a wide range of possibilities.

U.S. military RDT&E spending of about $35 billion a year remains three times the size of the rest of NATO's RDT&E spending combined. It is also roughly the size of either Russia's or China's entire defense budget. Defense RDT&E remains roughly equal in dollar magnitude to all other federally funded R&D in the country combined. It is also greater than average U.S. RDT&E spending in the 1950s, 1960s, and 1970s.[25] As shown in figure 7-4, it will remain relatively high by those decades' standards even at its intended 2005 real level of $31 billion (in constant 2000 dollars).

Of the scheduled decline in RDT&E spending between now and 2005, $3 billion is due to the anticipated completion of advanced development of the F-22, F/A-18 E/F, V-22, and a new attack submarine. As such, it does not represent a major lessening of the country's commitment to basic military research and innovation.[26] Some of the remaining $1 billion in reductions should be achievable by downsizing a DoD laboratory and weapons testing infrastructure of eighty-six facilities that has been spared much of the scrutiny of the base closure process so far but appears ripe for consolidation. Its capacity will remain some 35 percent greater than estimated needs even after all base closure rounds of the 1990s are fully completed.[27] In some

25. International Institute for Strategic Studies, *The Military Balance 1996/97* (Oxford University Press, 1996), p. 40; Office of the Under Secretary of Defense (Comptroller), *National Defense Budget Estimates for FY 1998* (Department of Defense, 1997), pp. 110–11.

26. Congressional Budget Office, *Reducing the Deficit: Spending and Revenue Options* (March 1997), pp. 26–27; Lane Pierrot, *A Look at Tomorrow's Tactical Air Forces* (Congressional Budget Office, January 1997), p. 24; Marine Corps briefing materials on V-22 Osprey, June 13, 1997.

27. General Accounting Office, *Defense Acquisition Infrastructure: Changes in RDT&E Laboratories and Centers*, GAO/NSIAD-96-221BR (September 1996), p. 2; "Pentagon Considers Restarting 'Vision 21' Lab Consolidation Effort," *Inside the Pentagon*, August 7, 1997, p. 1; Mark Walsh, "Congressional Hurdles Cloud Pentagon Lab Closure Plan," *Defense News*, June 9, 1997, p. 52.

Figure 7-4. *U.S. Defense Research, Development, Test, and Evaluation Spending, 1945–2005*

Billions of constant 2000 dollars

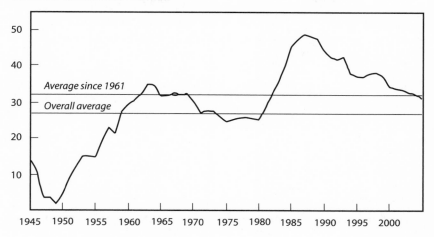

Source: Office of the Under Secretary of Defense (Comptroller), *National Defense Budget Estimate for FY00* (Department of Defense, March 1999), pp. 106–11.

cases, laboratories might be consolidated by reducing service-specific labs and establishing new joint centers, colocated when possible with testing facilities, that might be run by the office of the secretary of defense (or possibly a lead service for a given area of warfare such as air, sea, cyberspace, or space).[28]

Still, the scheduled cuts in RDT&E funding go a bit too far. Even though the RDT&E infrastructure remains too large, it is also aging and in need of refurbishment in a number of locations.[29] Nor does it make sense to keep reducing the real funding for RDT&E activities known as basic research, applied research, and advanced technology development (see figures 7-5 and 7-6). These three RDT&E accounts total only $7 billion. Demonstration and validation, engineering and manufacturing development, and operational systems development, all of which relate to specific weapons programs rather than basic technology development, account for the vast bulk of funds in the overall account. Restoring at least $500 million back to the more basic research activities of the RDT&E program would be well advised.

28. Robert Glen and others, "Losing the Technological Edge," Southwest Defense Alliance, 1999.

29. Philip E. Coyle, "Operational Test and Evaluation Overview for the Defense Science Board Task Force on Test and Evaluation" (Department of Defense, May 26, 1998).

Figure 7-5. *U.S. Department of Defense Basic Research Budget, 1972–2000ᵃ*

Billions of constant 2000 dollars

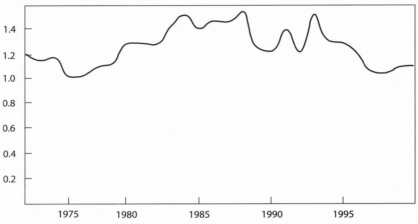

Sources: Statement by the Director of Defense Research and Engineering, 93 Cong. 1 sess., 1973, and the same publication for other years; "The FY1982 Department of Defense Program for Research, Development, Test and Acquisition," statement by the Honorable William J. Perry, Under Secretary of Defense, Research and Engineering, 97 Cong. 1 sess., 1981, and the same publication for other years; Office of the Under Secretary of Defense (Comptroller), "RDT&E Programmes (R-1)," January 1987, and the same publication for other years.

a. In FY98, the category of research was redefined as basic research.

Some of these funds might be used to increase funding for so-called advanced-concept technology demonstrations, which were funded at nearly $1 billion a year in the mid-1990s but have since been cut back to only about $100 million annually despite the fact that such funding produced several quick-turnaround capabilities of use in the war against Serbia.[30] Much of this reduction should be reversed.

Today the military services continue to make weapons, develop warfighting doctrine, and practice and train essentially on their own. Some joint exercises occur, but these focus primarily on ensuring immediate wartime readiness or other near-term goals. Thus, the National Defense Panel was right to promote greater focus on joint-service experimentation when weapons, tactics, and operational concepts are being developed and honed.[31]

30. Rachel Schmidt, *The Department of Defense's Advanced Concept Technology Demonstrations* (Congressional Budget Office, 1998), pp. 1–8; Daniel G. Dupont, "Citing Kosovo Success, DoD Fights for Advanced Concept Tech Demos," *Inside the Pentagon,* July 15, 1999, pp. 5–6.

31. National Defense Panel, *Transforming Defense: National Security in the 21st Century* (Arlingon, Va.: December 1997), pp. 68–73.

Figure 7-6. *U.S. Department of Defense Applied Research Budget, 1972–2000[a]*

Billions of constant 2000 dollars

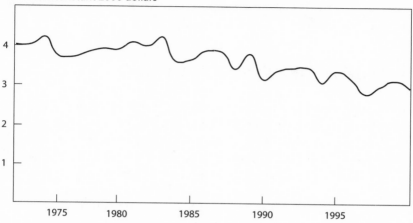

Sources: See figure 7-5.

a. In FY98, the category of exploratory development was redefined as applied research.

Accordingly, my alternative option adds $250 million a year to fund the recommended initiative. This amount corresponds to somewhat more than 1 percent of what the services currently spend on training.[32] It would allow several brigades, air squadrons, marine expeditionary units, and other comparably sized units from each service to undertake a number of fairly large-scale field exercises. By contrast, the Atlantic command, charged with conducting joint experimentation, had only about $20 million in 1999 to conduct field experiments using personnel from different services.[33] Admiral Harold Gehman, commander in chief of Atlantic command, considered that money about enough to do one major field experiment. He does not consider such experiments necessary in large numbers at present, given the ability to use simulators and other inexpensive techniques in many situations. However, he expects that within several years DoD may wish to carry out a dozen such activities annually, perhaps at a scale of $10 million or more per exercise. That seems an appropriate goal, commensurate with the intensity and scale of training carried out by the individual services at their

32. The military spends about $20 billion a year on operations and deployments. See William J. Perry, *Annual Report to the President and the Congress* (Department of Defense, February 1995), p. 42.

33. Robert Holzer, "Money Woes May Hamstring Pentagon Joint Experiments," *Defense News*, August 3, 1998, p. 1.

major training and test centers.[34] Former army chief of staff Dennis J. Reimer has laid out a similar philosophy of progressing from simulations into more experiments, perhaps making use of several services' major test and training ranges in the southwest United States.[35]

Economize on Major Weapons Platforms

With all these new needs to satisfy—totaling perhaps $7 billion a year for a decade—and a Pentagon already underfunded for its existing program and force, how is DoD to make ends meet fiscally? In terms of technology and military innovation, the answer is to alter the Pentagon's philosophy toward buying major weapons platforms. Rather than replacing most existing weapons systems with next-generation platforms as it now intends (see table 7-1), the Pentagon should take a high-low mix approach, generally purchasing only modest numbers of highly advanced systems. To satisfy the remainder of its requirements, it should either refurbish aging weapons or replace them with existing technologies. With this philosophy, it should be able to reduce its typical annual procurement funding requirement by at least $10 billion relative to current plans.[36]

In some cases, the United States could adopt this high-low approach by buying high-performance platforms for a one-war capability. That would provide a hedge against a country like China that may someday become more of a serious military rival to the United States. The rest of the force structure would be filled out with less advanced, less expensive equipment. For example, rather than replace virtually the entire tactical combat aircraft fleet with F-22 Raptors and joint strike fighters, the United States could buy modest numbers of these systems to constitute "silver bullet" fleets. It would then buy more F-15s and F-16s, modified only slightly (if at all) from the types of aircraft now in the U.S. military arsenal, to round out the force structure. Wielding improved munitions and supported by a greatly improved reconnaissance, computing, and communications infrastructure, these aircraft would be more capable than today's—and than aircraft adver-

---

34. "Gehman: First Experiment to Test Advanced Computing," *C4I News*, October 8, 1998, p. 1; Robert Holzer, "Modeling Holds Key to Joint Testing," *Defense News*, September 7, 1998, pp. 46–48.

35. Dennis J. Reimer, "Leaping Ahead to the 21st Century," *Joint Forces Quarterly*, no. 17 (Autumn–Winter 1997–98), pp. 20–24.

36. For more detail, see O'Hanlon, *How to Be a Cheap Hawk*, pp. 50–51, 98–136.

Table 7-1. *Major Pentagon Acquisition Programs, August 2, 1999*[a]

Weapon system	Total program cost (billions of dollars)	Planned quantity
Army		
ABRAMS upgrade	8,092.6	1,555
AFATDS (Advanced Field Artillery Tactical Data System)	1,353.2	6,454
ASAS (All Source Analysis System)	1,202.5	...
ATACMS/APAM (Army Tactical Missile System)	2,328.3	2,317
ATACMS/BAT (Army Tactical Missile System/Brilliant Anti-Armor Submunition)	6,252.6	21,454
ATIRCM/CMWS (Advance Threat Infrared Counter-measures/Common Missile Warning System)	2,996.3	2,590
Black Hawk (UH-60L)	5,101.8	646
Bradley FVS (Fighting Vehicle System) upgrade	4,065.3	1,109
Chem demil	14,586.9	15
Comanche (RAH-66)	8,168.5	8
Crusader (AFAS/FARV)	2,905.0	9
CSSCS (Combat Service Support Control System)	395.1	3,196
FAAD C2I BLK-II-I	1,975.8	...
FMTV (Family of Medium Tactical Vehicles)	18,383.6	86,916
ICH (Improved Cargo Helicopter)	3,026.1	302
Javelin	4,130.5	27,004
JSTARS GSM (Ground Station Module)	1,433.7	139
Longbow Apache	8,696.6	850
Longbow Hellfire	2,508.9	12,905
MCS (Maneuver Control System)	1,648.1	8,088
MLRS Upgrade (Multiple-Launch Rocket System)	4,933.9	63,023
SADARM (Search and Destroy Armor)	2,570.1	50,189
SINCGARS (Single-Channel Ground and Airborne Radio System)	3,843.3	270,384
SMART-T (Secure Mobile Anti-jam Reliable Tactical Terminal)	764.2	320
Subtotal	111,362.9	...
Navy		
AAAV (Advanced Amphibious Assault Vehicle)	934.1	...
AIM-9X	3,093.1	10,129
AV-8B Remanufacture	2,117.4	72
CEC (Cooperative Engagement Capability)	3,594.6	231
CH-60S	3,185.8	166
CVN-76	4,538.9	1
CVN-77	5,210.3	1

Table 7-1. *Continued*

Weapon system	Total program cost (billions of dollars)	Planned quantity
DD 21	3,191.1	...
DDG 51 (destroyer)	53,965.2	57
E-2C reproduction (early warning and control aircraft), Hawkeye	3,130.2	36
F/A-18 E/F Super Hornet	47,007.8	548
JSOW (Joint Standoff Weapon)	7,285.3	23,924
LHD 1 (amphibious assault ship—U.S.S. *Wasp*)	7,826.3	7
LPD 17 (LX) (amphibious assault ship)	9,881.4	12
MHC 51 (minehunter, coastal—Osprey class)	1,751.2	12
MIDS-LVT (Multifunctional Information Distribution System–Low Volume Terminal)	1,352.1	2,438
NESP (Navy EHF Satcom Program)	2,084.9	359
SH-60R (LAMPS-BLK II)	4,963.7	188
SSN 21 (attack submarine) 21/AN/BSY-2	13,409.2	3
SSN 774 (NSSN)	65,152.9	30
STD MSL 2 (BLKS I-IV)	9,629.2	11,505
Strategic sealift	5,835.4	19
T-45TS	6,870.7	236
Tactical Tomahawk	1,863.4	...
Tomahawk TBIP	470.0	...
Trident II MSL	27,355.6	453
USMC H-1 upgrade	3,643.7	284
V-22 Osprey	36,220.3	458
Subtotal	335,563.8	...
Air Force		
ABL (Airborne Laser)	2,173.9	2
AMRAAM (Advanced Medium-Range Air-to-Air Missile)	10,359.2	10,917
AWACS (Airborne Warning and Control System) RSIP (E-3)	910.5	32
B-1B CMUP	2,117.0	287
C-130J	2,859.6	37
C-17A	44,913.6	134
DMSP (Defense Meteorological Satellite Program)	2,423.9	10
EELV (Evolved Expendable Launch Vehicle)	17,347.8	181
F-22 Raptor	62,740.1	341
GBS (Ground-Based Sensor)	457.7	493
JASSAM (Joint Air-to-Surface Standoff Missile)	2,006.1	2,461

Table 7-1. *Continued*

Weapon system	Total program cost (billions of dollars)	Planned quantity
JDAM (Joint Direct Attack Munition)	2,590.9	88,116
JPATS (Joint Primary Aircraft Training System)	3,905.3	712
JSIPS (Joint Service Imagery Processing System)	640.7	35
JSTARS (Joint Surveillance Target Attack Radar System)	8,351.2	15
MILSTAR satellites	...	6
Minuteman III (GRP)	2,403.6	652
Minuteman III (PRP)	2,176.4	607
NAS	787.1	75
NAVSTAR satellite	10,151.4	115
NAVSTAR user equipment	6,714.3	233,120
SBIRS (Space-Based Infrared System)	4,002.3	5
SFW (Sensor-Fuzed Weapon)	1,920.9	4,920
Titan IV	17,567.9	39
Subtotal	209,521.4	...
DoD-wide		
JSF (Joint Strike Fighter)	23,362.5	...
Navy TBMD	6,709.6	...
NMD (National Missile Defense)	8,808.4	...
NPOESS (National Polar-Orbiting Operational Environmental Satellite System)	4,929.3	...
Patriot PAC-3	7,778.3	...
THAAD (Theater High-Altitude Air Defense)	8,692.8	...
Subtotal	60,280.9	...
Total[b]	706,935.6	

Source: Department of Defense, "News Release: Selected Acquisition Reports," August 19, 1998, August 2, 1999, and March 31, 1999.

a. Total estimates have decreased by $10,212.4 million since December 1997 for the following programs: AFATDS, ASAS, CSSCS, FAAD C2I, MHC 51, Tomahawk TBIP, DMSP, and JSIPS. List includes only those systems and quantities that have been formally approved.

b. Subtotals may not add to total due to rounding and the addition of the navy theater-wide missile system in June 1999.

saries will be able to purchase in significant numbers—even if their airframes and engines remained unchanged (see figure 7-7).

A similar high-low philosophy would also guide purchases of the V-22 Osprey tilt-rotor aircraft and the navy's F/A-18 E/F. Rather than buying around 500 of each of these systems, the military might buy only one-third to one-half as many. For the V-22, that would still provide enough capa-

Figure 7-7. Quality of Fighter Aircraft, 1995, 2005, and 2015[a]

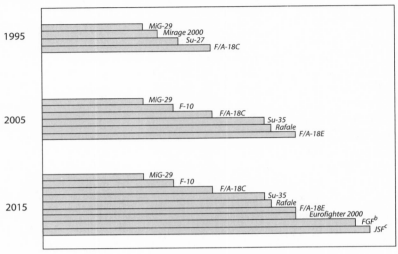

Overall effectiveness (notional units)

Source: Adapted from Office of Naval Intelligence, *Worldwide Challenges to Naval Strike Warfare* (January 1996), p. 35.

a. MiG and Su aircraft are Russian, the F/A-18 and JSF are American, the Rafale and Mirage are French, and the Eurofighter is being built by a multinational West European consortium.

b. Russian Future-Generation Fighter.

c. U.S. Joint Strike Fighter.

bility for longer-range counterterrorism or search and rescue missions, where the Osprey's speed and range make a great difference in mission effectiveness. For the Super Hornet, that would provide enough aircraft for targets located at greater distances, or for targets situated near dense air defenses, where the F/A-18 E/F's range and stealthiness would most come in handy. Otherwise, the marines and navy could replace current systems with transport helicopters and with new F/A-18 C/D's.

Consider as well the army's Comanche helicopters, designed both for light attack and for escorting Apache heavy-attack helicopters. Given trends in unmanned aerial vehicles and DoD's information networks, a manned scout platform may no longer be needed for the Apache. The Apache demonstrated in Desert Storm that it was indeed capable of flying without escort at quite modest risk.[37]

Far from weakening future U.S. military forces, this approach could well strengthen them relative to the plans contained in the QDR. It would take

37. O'Hanlon, *How to Be a Cheap Hawk*, pp. 48–97.

advantage of the most promising new technologies and put emphasis on keeping weapons reliable, safe, and serviceable, rather than try to modernize virtually all weaponry when that is neither necessary nor affordable. The QDR's modernization agenda is a bit of a high-risk gamble with the basic health and fighting readiness of the U.S. military. If the real defense spending increases that it presupposes do not materialize, the Pentagon's natural reaction will probably be to slow down procurement programs year to year, always hoping that the next year's budget will allow it to compensate for time lost. As a consequence, existing equipment will be retained longer and become less dependable. Moreover, the problem feeds on itself; when fewer weapons are purchased than planned, the unit costs of those built tend to go up since inefficient use is made of production facilities.[38]

Conclusion

The United States can devise a more affordable military modernization strategy by emphasizing the first major premise of RMA proponents as laid out in chapter 1: that electronics and computing are indeed offering remarkable new capabilities at modest cost. By focusing on this specific technical foundation, and emphasizing R&D and experimentation as well, the United States can carefully explore the hypothesis that a contemporary revolution in military affairs is possible without embarking on an ill-advised near-term transformation strategy for the core of its military that would have huge budgetary and opportunity costs.

The United States should modernize its armed forces when that can be done wisely and economically in the years ahead. It should also pay particular heed to addressing the vulnerabilities in its defense capabilities that others may seek to exploit in the future. It is right to try to preserve an overwhelming military dominance to maximize deterrence and to limit the expected number of American casualties in any future conflicts. But the United States also needs a sense of perspective about how good its military is today. It further needs to remember that, as it demonstrated during the cold war, fashioning a successful security strategy requires more than military technology.

38. See R. William Thomas, *Effects of Weapons Procurement Stretch-Outs on Costs and Schedules* (Congressional Budget Office, 1987), pp. ix–xiii, 31.

CHAPTER EIGHT

Conclusion

Is a revolution in military affairs (RMA) achievable at the turn of the twenty-first century, and if so does it necessitate a radical change in U.S. military equipment, combat structures, and warfighting doctrine? Or can the United States continue to make security policy, and arrange Pentagon budgetary priorities, in a more continuous and evolutionary way?

The RMA debate is performing the nation a service by forcing it to confront these questions directly. Victorious military powers tend to settle into strategic complacency, allowing their influence and security to decline over time; military services tend to be conservative and to resist bold innovation. Explicitly considering various options for the future U.S. armed forces, and thinking in bold terms rather than simply debating how many existing fighter wings, divisions, and aircraft carriers we should maintain, reduces the risk that these dangers will befall the United States.

That said, the contemporary revolution in military affairs hypothesis is unconvincing, and the technological basis for a radical RMA transformation of the U.S. armed forces and U.S. security policy is unsubstantiated. If pursued as many advocate, moreover, it could be counterproductive for the country's long-term security interests by reducing funding for various forms of global military engagement, preventive defense, and old-fashioned deterrence.[1] Even if a revolution in military affairs does happen sometime in the early part of the twenty-first century, its scope will probably be more lim-

---

1. On preventive defense, see Ashton B. Carter and William J. Perry, *Preventive Defense* (Brookings, 1999).

ited, and the time scale over which it is carried out longer, than most proponents now believe. And it may well result from a sustained period of rapid evolution in military technology and doctrine—already the norm in U.S. defense policy for decades—rather than an abrupt remaking of the armed forces.

Technological modernization is important, but there is little reason to think that such modernization is in serious jeopardy provided the United States modestly increases the amount of money now spent on military acquisition—that is, the sum of procurement with research, development, testing, and evaluation (RDT&E) spending. Current spending levels of nearly $90 billion need to increase by some $10 billion to $15 billion a year to sustain RDT&E, increase joint-service experimentation, replace aging equipment, and allow targeted modernizations. That acquisition budget would be roughly twice as large as what any other country spends on its entire military establishment. It would not be enough to fully fund the current Pentagon modernization agenda—doing so could require $120 billion in annual funding for acquisition accounts—but it should suffice for the most truly pressing needs.

These conclusions are not necessarily timeless. Within a few decades, biological agents, directed-energy weapons, or other new defense technologies may come to dominate warfare. At such a point, preexisting weaponry may become obsolete, and a legitimate revolution in military affairs may occur. At present, such possibilities remain more within the realm of science fiction—or, at best, the realm of basic scientific research—than of defense planning. This argues for maintaining vigorous RDT&E activities, not for rapidly embarking upon a transformation strategy to restructure and reequip the bulk of the U.S. military today.

To put it differently, even if a contemporary revolution in military affairs may eventually be possible, it does not appear within reach today. The relevant historical analogue may be the situation confronting the United States and other major military powers in the 1920s (though progress in most major military platforms is much less rapid today than it was then). The current era, however, is probably not akin to the situation of the 1930s and 1940s, when operational concepts such as carrier warfare, blitzkrieg, and radar-directed air defense as well as nuclear weapons were developed. In the earlier era, some basic new technologies like radio and radar were being discovered, airplanes were becoming robust enough to fly off ships, and vehicles were becoming dependable enough to make mechanized warfare a possibility. But operational concepts for using these systems were in their very early stages, and the technologies undergirding them remained unrefined.

Given this prognosis, robust research, experimentation, and prototyping make more sense than an attempt to radically transform military hardware, organizations, or warfighting concepts. Fortunately, given the extent to which the United States has institutionalized vigorous military research and development (R&D) even in peacetime, adoption of such a strategy does not require major changes in most areas of the Pentagon's budget.

The reasons for skepticism about an imminent RMA are straightforward. As this study has emphasized, most trends in defense technology are not as impressive as RMA proponents argue. It is true that an electronics and computer revolution is occurring—with important implications for warfare—but in other important areas progress is far slower. The propulsion systems and the basic aerodynamics and hydrodynamics of ships, planes, rockets, and ground vehicles are generally changing at modest rates. The explosive power of conventional ordnance per unit weight is advancing only modestly. Armor is becoming lighter and stronger but in increments of 10 and 20 percent improvement per decade, not 50 and 75 percent.

Equally important obstacles stand in the way of radically improving sensors, despite the goal of RMA proponents of achieving dominant battlespace knowledge. Sensors will progress in important ways, largely because they will become smaller, making it possible to place them in greater numbers on a wider array of manned and unmanned platforms. Their abilities to see through substances like water, metal, wood, soil, and buildings will remain seriously limited by practical engineering considerations, however, and even more so by the laws of physics. These physical limitations will be exploited by future enemies, who will increasingly find ways to make military vehicles resemble civilian vehicles in external shape and in their radar and infrared signatures and who will continue to have little trouble hiding small arms from sensors in most types of tactical settings.

Communications systems are improving more rapidly. But they, too, will have limitations. They could be vulnerable to various types of enemy attack, including possible enemy use of high-altitude nuclear explosions that would do much more harm to the advanced electronics of the U.S. military than to the simpler weaponry operated by most of its foes. Most such vulnerabilities can be mitigated by building redundant systems and hardening electronics. Yet so far the United States is showing only limited concern about this potential Achilles' heel of its future military, skimping on radiation hardening and otherwise assuming too blithely that future enemies will simply not dare to use high-altitude nuclear bursts against it.

The situation with communications systems is emblematic of a broader reality. Not just the United States and its allies but also enemy forces will make use of advanced military technologies wherever possible. As noted, they will certainly make maximum use of countermeasures, decoys, cover, and other impediments to sophisticated U.S. sensors. They will also acquire relatively inexpensive systems, such as night-vision capabilities and laser range finders, through the global arms market. Adversaries will acquire advanced antiship and antiaircraft missiles, mines, and cruise and ballistic missiles as well.

That said, computers and communications systems do offer the U.S. military great promise. It is well placed to exploit these capabilities, given that it is already far ahead in such spheres of defense technology and that it has the most computer-proficient economy in the world to draw upon for technology and human expertise. These raw ingredients do not ensure success, but they provide great potential for it.

U.S. military forces face major challenges above and beyond the purely technological ones as well. They must avoid service parochialism and foster jointness while developing information networks for warfare. They must do as much as possible to enhance and ensure allied interoperability. They must avoid complacency about the invulnerability of their communications and electronics systems and provide adequate resources to ensure radiation hardening and systems redundancy. More broadly, they must resist the temptation to keep spending most modernization dollars on major weapons platforms. That tendency jeopardizes the availability of funds for computers, communications systems, and munitions as well as for more mundane necessities like improved chemical protection gear, greater physical hardening of overseas bases, and increased numbers of fast sealift ships. If these kinds of pitfalls can be avoided, the U.S. armed forces can make great strides—not only in absolute terms but also vis-à-vis the likely competition.

Several other important policy implications follow as corollaries from the aforementioned prognostications. Since modernizing electronics and munitions is relatively inexpensive, it is not beyond the fiscal means of most U.S. allies to keep up with key trends in defense technology and military innovation—provided they properly order their defense priorities. The common notion that the so-called revolution in military affairs will necessarily price them out of the market, widening the gap between their armed forces and those of the United States, is thus incorrect. Second, deploying military forces abroad will remain time consuming and difficult. Technology may offer some ways to make combat units lighter and faster, but trends in engines, armored vehicles, ships, planes, rockets, and explosives do not offer any realistic promise of being able to employ military force rapidly and decisively from long range for most purposes. Third, constraints on sensor technology mean that it will not be possible to verifiably eliminate nuclear weapons from the face of the earth, even were such a goal deemed desirable and politically achievable, for the foreseeable future.

Whatever their capabilities, new sensors, communications systems, computers, and weapons are likely to make much more difference in some spheres of combat than in others. The main point is simple: warfare involving large objects on open terrain is likely to favor the high-technology force and hence the United States—perhaps even more than it already does today. This observation is not without importance; it applies, for example, to a potential Chinese invasion of Taiwan and to a potential reprise of an Iraqi attack against Kuwait or Saudi Arabia. In broader terms, however, there is little reason to think that Desert Storm was a harbinger of wars to come. Cross-border aggression has been becoming less prevalent in the world for decades, and that trend is likely to continue or at least not be reversed.[2]

Moreover, any future aggressors are likely to be more wiley than Saddam. They may frame their potential victims, confusing the international community about which side began a given conflict long enough to seize key objectives. They may strike their victims quickly first, disguising mobilization efforts as maneuvers, and then offer compromise solutions to discourage major military responses from the United States and the world community more broadly. They may also make explicit or implicit threats to use weapons of mass destruction should the United States and like-minded countries try to overturn their conquests.

Under such conditions, high-technology forces are unlikely to prove as dominant as they were in Desert Storm, even with next-generation weaponry. They will still have to evict an aggressor from the land it has conquered in wars where U.S. responses are delayed by political hesitation, geographic distance, or some combination of the two. Even in places where the United States and other countries respond fast enough to defend an ally or neutral country from attack, they will need to uproot enemy forces in situations where it becomes necessary to overthrow an adversary's government and occupy its country. Such conditions could very easily apply in a place like Iraq or North Korea, where it is doubtful that the United States would be

---

2. Yahya Sadowski, *The Myth of Global Chaos* (Brookings, 1998), pp. 87–88.

willing to leave current regimes in power should they again undertake large-scale wars of aggression against American friends and allies.

These conclusions mean, in turn, that U.S. forces will often need to prevail against foes whose troops may well be interspersed among civilian populations and in combat settings where even relatively unsophisticated enemy units will have opportunities to ambush American troops or booby-trap and mine their likely paths of advance. In such situations, U.S. forces will not have the luxury of attacking an enemy while staying safely out of harm's way.

The same basic assessment pertains to many other settings in which the United States is likely to use military force in the future. Whether conducting peace operations, searching for drugs or nuclear materials, attempting to arrest war criminals, or tracking down terrorists, the U.S. military will need to operate in complex terrain and cope with adversarial forces that may be difficult to identify. That means enemy units will often have the chance to get in the first shot against American troops.

New tools and weapons will change many of the tactical specifics of these scenarios, but not their basic nature. Environments such as forests and cities will remain significant obstacles to the effectiveness of high-technology systems. U.S. and other advanced forces would still enjoy important military advantages in places such as the Korean peninsula, the hills of the Balkans, or the cities of countries such as Somalia. But they will depend as much on their basic military skills as on technology and will often not be able to count on detecting and firing at the enemy before coming within range of that adversary's own weapons.

Despite these limitations and caveats, not all the news is bad. Rather than necessitating a wholesale transformation of the U.S. armed forces, current technological trends allow the United States to pursue an impressive military modernization strategy relatively inexpensively, and without the need to curtail its other security activities around the globe. At the turn of the twenty-first century, that approach makes sense in broader foreign policy terms as well. The potential ability of the United States to help consolidate a revolution in geostrategic affairs—in which most of the world's major industrial powers are democratic, prosperous, allied with each other, lacking a major strategic foe, and gradually extending their club of membership to other countries—is even more historic, and more important, than its purported ability to again revolutionize warfare.

Index

Accidental death in combat, 125, 126
Acoustic sensors, 44, 50, 67, 148–49
Acoustic weapons, 91, 92
Advanced concept technology demonstrations, 184
Air forces, 29; aircraft materials technology, 70; Battle of Britain, 23; future deployment patterns, 145–47; future of infantry combat, 113, 148; global reach, global power concept, 14–15, 140–41; importance of U.S. superiority in, 146, 148; in heavy combat situations, 119–28; in peacekeeping and humanitarian interventions, 128–29, 131–32; infrared navigation and targeting systems, 36; naval, 23; performance in Persian Gulf War, 10, 14; propulsion technology, 69–70; prospects for technological development, 139; recent technological advances, 27, 69–73; reductions in force, 152–53; short-range combat aircraft, 148; stealth technologies, 71; surveillance capabilities, 34; targeting technology, 175–76; transport aircraft, 72–74, 145, 178; weapons platform

development, 186–90. *See also* Rocket technology
Airborne warning and control system (AWACS), 48
Aircraft carriers, 3, 21, 23; technical development, 70–71
Airlift, 73–74, 145, 151, 178
Allied forces, 155–58
Ambush, 117, 126
Anti-Ballistic Missile Treaty, 99–100
Apache helicopter, 190
Arab-Israeli War (*1973*), 9, 27
Argentina, 138
Armored gun system, 178
Armored vehicles: antiarmor weapons, 81–82; camouflage, 124–25; future of infantry combat, 113; lightweight, 16, 130, 134; mine-resistant, 178–79; Persian Gulf War, 10; prospects for peacekeeping and humanitarian missions, 129–30; prospects for technological development, 123, 139–40; radar countermeasures, 46–47; ships, 134–35, 137–38. *See also* Tanks
Arms sales, 136
Army, U.S.: communications technology,

53–54; global reach, global power concept, 16; tactical missile system, 87–88

Arsenal ships, 137–38, 147–48

Artillery, 21; prospects for technological development, 122; targeting technology, 114

Australia, 155

Automatic target recognition, 57, 61–62

AWACS. *See* Airborne warning and control system

B-2 bomber, 146–47

Base closures, 182

Battery power, 84

Biddle, Steven, 127

Bin Laden, Osama, 114–16

Biological warfare: delivery technology, 101–02; domestic defense against, 181; nuclear deterrence of, 163–64; prospects for technological development, 105; research and development, 94–95; sensor technology, 36, 43–44, 51, 164

Bistatic radars, 40–41, 110

Blimps, 74

Blitzkrieg, 3, 8, 21, 23

Bosnia, 144

Brilliant antitank weapon, 88, 122, 173

Britain, 23, 71, 133–34, 138, 157

Budget issues, 24–31, 168–91

Butler, Lee, 161

C4 technologies, 13, 176

Camouflage, 35, 36; artillery, 114

Casualties: accidental death in combat, 125, 126; Operation Desert Storm, 125–26; public support of military operations and, 7

Chemical warfare: domestic defense against, 181; research and development, 94; sensor technology, 43

China, 60, 136, 168, 196

Chromatography, 43

Civilian role in RMAs, 23, 26

Clausewitz, Carl von, 8

Clinton administration, 100, 177, 180–81

Cohen, William, 25, 155, 179

Comanche helicopter, 72, 190

Combat, 5, 142; army dispersion patterns, 121; battlefield communications technology, 52–54, 116; dominant battlespace knowledge, 11, 13–14; heavy weaponry conflicts, 119–28; infantry engagements, 112–19; peacekeeping and humanitarian interventions, 128; planning, 179–80; prospects for battlefield information technology, 108–12, 115–17; relative importance of technology, 196; technology-thwarting tactics, 117–18

Communications technology: allied interoperability, 155–56; antisurveillance measures, 49–50; bandwidth limitations, 58; battlefield applications, 52–54, 116; countermeasures, 58–61, 111; encryption technology, 50; fiberoptic, 49, 54; microwave radar, 40; near-term technology development goals, 174–76; prospects for technical development, 52–55, 64–66, 116, 194; RMA premise, 106; RMA shopping list, 157; vulnerability, 124

Computer technology: alliance compatibility, 155–56; near-term technology development goals, 173–76; prospects for RMA, 141–42; prospects for technical advancement, 56–57, 64, 66, 109, 195; rate of technological advancement, 30, 55–56, 66; RMA premise, 12–13, 106, 107; tactical Internet capabilities, 175; vulnerability of military systems, 62–63; weapons simulations and modeling, 26, 63–64

Cost of technology: battlefield communications system, 53; innovation,

156–57; military satellite technology, 39–40; multinational collaborations, 156–57; near-term computer and electronics investments, 175; recommended investment areas, 172–73; RMA shopping list, 157–59; RMA spending proposals, 169–70; rockets, 77; to redress U.S. vulnerabilities, 176–77; transport capacity, 178

Cox, Christopher, 60

Coyle, Philip, 99, 103

Davis, Tom, 176

Decoys: detection, 35, 46; theater missile defense, 100–01

Defense spending: computer and electronics allocations, 173–76, 191; cost of RMA implementation, 29; current U.S., 153; domestic biochemical attack, 181; future needs, 193; future policy challenges, 6, 171–72; future prospects, 170–71, 182; international comparison, 153, 171–72, 182, 195; large-scale field exercises, 181–86; mine warfare, 179; modernization effort, 4; national missile defense, 180–81; political context, 154; recommended investment, 172–73; research and development, 181–86; RMA allocation, 24–25, 168, 169–70, 191; weapons platform development, 186

Defensive systems: against biological weapons, 95; against chemical weapons, 94; against radio-frequency weapons, 111, 174–75; fixed battlefield assets, 124; missile, 96–104; RMA shopping list, 158; underground construction, 90, 111

Department of Defense (DoD): modeling and simulation capabilities, 26; modernization budget, 169–70; RMA policy, 18–20, 25; RMA policymaking role, 26

Deterrence, 152, 163–66; rapid troop deployment for, 177

Diamond, Jared, 22

Discoverer-2 programs, 39–40

DoD. See Department of Defense

Dominant battlespace knowledge, 11, 13–14, 67

Dominant maneuver, 19, 140

Electric guns, 83, 89

Electromagnetic pulse, 59–60, 111, 174–75

Encryption technology, 50

Exocet missile, 86, 133–34, 138

Explosives technology, 90, 111

F-15 aircraft, 186

F-16 aircraft, 186

F-22 aircraft, 146, 182, 186

F/A-18 E/F, 189–90

Falklands War, 133–34, 138–39

FBCB2, 175

Fiber-optic technology, 49, 54

Focused logistics, 19, 140

Fogleman, Ronald R., 13

Follow-on forces attack, 9

Force structure: global reach, global power concept, 15; RMA and change in, 20

Foreign affairs: implications of RMA, 143, 154–55; overseas basing, 144–53; security spending, 171–72

France, 23

Freedman, Lawrence, 141–42, 154–55

Fuels, 84–85, 134

Full-dimensional protection, 19, 140

Fusion technology, 93

Gansler, Jacques, 155

Garwin, Richard, 181

GBU-28, 90

Gehman, Harold, 185

Germany, 152

Global positioning systems, 53, 54–55, 174, 176

Global reach, global power concept, 12, 14–16

Ground forces: forward deployment levels, 152; future battlefield challenges, 149–50; future forward-deployment patterns, 145; future significance, 148; global reach, global power concept, 15–16

Guerrilla warfare, 22, 132–33

Guided weaponry, 1, 14, 27; access of U.S. adversaries to, 76; automatic target recognition, 57, 61–62; developments in naval combat, 133–34, 136–37; historical development, 9; infrared sensing, 35; missile technology, 85–87, 110; national missile defense, 180–81; of U.S. adversaries, 102; prospects for technological development, 121–22; radar imaging, 41; RMA shopping list, 158

Heavy combat, 119–28

Helicopters, 27, 113, 190; stealth technology, 71; transport, 73; vulnerability, 118–19

High-altitude nuclear bursts, 58–59, 174, 194

High-speed antiradiation missile, 86

Historical development, 1; advances achieved without RMA, 27–29; army dispersion patterns, 121; guided missile technology, 85–86; historical correlations with present, 5, 193; misperceptions of military technology, 29; modern military technology, 9, 27; naval weaponry, 133–34; past RMAs, 20–24; radar, 38; significance of RMA, 3–4, 7, 17; stealth technologies, 71; technological basis of RMAs, 5–6

Hitler, Adolf, 23

Hydrogen-fuel-cell technology, 84–85

Hydrogen gas guns, 89

Hypersonic aircraft, 78–79

Identification-friend-or-foe, 126

Infantry engagements, 112–19; peacekeeping and humanitarian interventions, 128; prospects for change, 139

Information technology: allied interoperability, 155–56; DoD goals, 19, 140; DoD modeling and simulation capabilities, 26; dominant battlespace knowledge, 11, 13–14; prospects for RMA, 141–42; RMA premise, 2, 12–13, 106, 107, 108–12; weapons targeting, 114–16, 121–22, 131–32. *See also* Communications technology

Infrared sensors, 34–35, 36; limitations, 45–46

Intelligence, surveillance and reconnaissance, 13; antisatellite weapons, 137; biological agents, 164–65; current capability, 120; for weapons targeting, 114–16, 131–32; future needs, 148; high-technology scout teams, 123–24; identification-friend-or-foe technology, 126; Mogadishu firefight, 118–19; near-term technology development needs, 176; peacekeeping and humanitarian missions, 131; radar imaging, 38, 39, 48; radio emission, 41, 49–50; radioactive materials, 44, 51, 67, 162–63; recent investments, 28; RMA shopping list, 157; visible light imaging capabilities, 34–38. *See also* Communications technology

Intercontinental ballistic missiles, 89; national missile defense, 97–100

Iran, 136, 168

Iraq, 168, 173, 196–97; air attacks (1998), 2, 27

Japan, 152

Joint direct attack munition bomb, 28, 86, 132

Joint Requirements Oversight Council, 176

Joint standoff weapon, 174

Joint strike fighter, 71

Joint surveillance and target attack radar system (JSTARS), 28, 39, 48, 120, 121–22, 132, 157, 176

Joint Vision 2010, 2, 19, 67, 69, 104, 107, 140

Jones, James, 5

KH-*11* spy satellite, 34, 39

Korean peninsula, 96, 144, 152, 168, 173, 177, 196–97

Kosovo, 2, 7, 28, 120, 128–29, 150, 177

Krepinevich, Andrew, 6, 20, 70

Krupp, Dennis, 179

Lacrosse satellite, 38, 39, 40

Landing-craft air cushion vehicles, 79

LANTIRN navigation system, 36

Laser technology: bomb and missile guidance, 9, 86, 87, 88; missile defense systems, 103; sensors, 34, 36–37

Lidar technology, 34, 36–37, 88

Liddell Hart, B. H., 9, 118

Logistical support, 123, 149–50; fuel supply, 84; global reach, global power concept, 15–16

Low-cost autonomous attack system, 88

M*1* Abrams tank, 81, 82

Marine Corps, U.S., 157; global reach, global power concept, 15–16

Mark, Hans, 78

Marshall, Andrew, 5

Materials technology, 113; tank design, 81, 82–83

Mechanical technologies: prospects for advancement, 3; rate of technological advancement, 68–69; significance of, 68

Metallurgy, 21; aircraft materials, 70; tank design, 81, 82

Microwave imaging, 40

MILSTAR system, 60

Mine warfare, 114; mine-resistant vehicles, 178–79; mine hunting and minesweeping, 135, 179; technology development needs, 178–79

Missile defense, 96–108, 180–81

Moffett, William, 23

Mogadishu firefight. *See* Somalia peacekeeping action

Moore's law, 30, 56–57

Moore, Gordon, 56–57

Multinational military operations, 25, 143, 166–67; defense spending, 171–72, 195; for peacekeeping and humanitarian interventions, 159–60; overseas basing, 144–53, 166; power imbalances among allies, 153–55; research and development, 156–57; RMA shopping list, 157–59

MV-*22* Osprey, 15

Napoleonic era, 20, 21, 22

National Defense Panel, 18–19, 20, 25, 30, 69, 145, 184

NATO, 154, 155, 156, 157

Naval combat, 133–39

Naval forces, 20–21; aircraft carrier design, 70–71; airpower, 23, 70–71; antiship missiles, 86, 135; arsenal ships, 137–38, 147–48; deployment speed, 134; Falklands War, 138; military sealift, 149, 150; mine warfare, 135, 179; propulsion technology, 79–80, 135; prospects for technological change, 134–38, 140; roll-on/roll-off vessels, 178; ship design and construction, 79–81, 110, 134–35; sonar detection, 42, 44; vulnerabilities, 133–34, 135–39, 140

Navigation systems, 36

Navy theater-wide (NTW) missile defense system, 97–99, 100
Night vision sensors, 36
Nonlethal weapons, 91–93
Nuclear Nonproliferation Treaty, 160–61
Nuclear weapons, 21; computer simulations and modeling, 63–64; current concerns, 161–62; deterrence value, 163–66; disarmament movement, 160–61; electromagnetic pulse effects, 59–60, 174, 194; high-altitude nuclear bursts, 58–59, 194; low-yield, 93; monitoring technology, 162–63; proliferation, 93–94; prospects for elimination, 143, 160–66; prospects for technological development, 93–94; seismic monitoring, 50–51

Ochmanek, David, 173
Ogarkov, Nikolai, 9
Okinawa, 144, 145
Operation Allied Force, 28, 74, 87, 155
Operation Desert Storm, 1, 10, 26, 49, 81, 86, 87, 125–28, 177; alternative scenarios, 126–28; casualties, 125–26; reconnaissance and surveillance, 39, 112; significance of, 27, 28. *See also* Persian Gulf War
Organizational change, 20
Overseas basing, 144–53, 177
Owens, William, 13, 176

Pakistan, 136
Panama invasion (*1989*), 1, 112
Patriot missile, 96, 97, 100
Peacekeeping and humanitarian interventions, 128–33; Kosovo case study, 131–33; military skills for, 128; multinational collaboration for, 159–60; prospects for technological future, 129–31, 133; role of air power in, 128–29, 131–32
Pentagon. *See* Department of Defense

Pentomic division, 29
Persian Gulf War, 7, 96; air force, 14; conventional forces, 10; significance of, 9–11; weapons performance data, 9–10. *See also* Operation Desert Storm
Policymaking: antisatellite weapons technology, 137; civilian role in RMAs, 23, 26; DoD and RMA, 18–20, 26; future challenges, 195; future security investments, 6; implications of RMA, 4, 25–26
Precision engagement, 19, 104, 140
Precision munitions, 173–74
Predictive modeling: DoD performance, 26; technology development, 32, 33
Propulsion: engine-disabling weapons, 91; ground vehicle, 83–85; microelectromechanical, 76; missiles, 85–86; prospects for technological development, 104, 107, 111; rocket technology, 76–79; ships, 79–80, 134, 135
Public opinion: casualties in conflict, 7; in defense policy formulation, 7–8

Quadrennial Defense Review (QDR), 18–19, 19, 20, 25, 146, 190–91

Radar technology, 34; antistealth, 41, 47; antiradar missiles, 86; bistatic, 40–41, 110; countermeasures, 46–47; current applications, 38–39; foliage-penetrating, 40, 49, 66–67, 117; for targeting of munitions, 131–32; historical development, 38; phased-array, 38, 39; prospects for advancement, 38–41, 66–67; ship detection, 79, 80–81; technical limitations, 46–50; wall-penetrating, 117
Radio emission detection, 41, 49–50, 115; antiradar missiles, 86–87
Radio-frequency weapons, 111, 132, 174–75

Radioactive materials detection, 44, 51, 67, 162–63
Reimer, Dennis J., 186
Research and development: antisatellite weapons technology, 137; battlefield sensors, 33, 42–43; biological weapons, 94–95; chemical weapons, 94; computer technology, 56–57; defense budget allocation, 4, 6, 191, 193; ground vehicle design, 83–85; guided missiles, 85–87; hypersonic aircraft, 78–79; joint-service activities, 184; large-scale field exercises, 185–86; missile defense systems, 97–99, 100–01; multinational efforts, 156–57; near-term needs, 181–86; need for, 1–2; policy issues, 4; predicting performance, 32, 33; radar and radio emission sensing, 38–41, 46–50; recent allocation, 28; recommendations, 6, 172–73; RMA allocation, 24; rocketry, 77–79; tank design, 82–83; transport aircraft, 73–74; unmanned aerial vehicles, 74–76; visible light sensing, 34–38, 45–46; weapons simulations and modeling, 26, 63–64
Revolution in military affairs (RMA): advances achieved without, 26–29; allied interoperability, 155–56; among U.S. adversaries, 3, 4, 12, 16–17, 29–30, 107, 195; as catalyst for innovation, 7–8, 192; as change in organization or operations, 20; civilian role, 23, 26; cost of implementing, 157–59, 169–70; costs of misperceiving, 29–30, 31; creation, 24; defense spending, 24–25; definition, 2, 11; DoD policy, 18–20, 25; dominant battlespace knowledge school, 11, 13–14; global reach, global power school, 12; high-intensity warfare, 125; historical examples, 20–24; implications for foreign relations, 154–55, 166–67; impli-

cations for overseas basing, 144–45, 151–53, 166; implications for U.S. military power, 2, 3; indications, 5, 29, 31, 32–33; information technology, 67, 106, 107, 108–12; likelihood, 108–12, 141–42, 192–93, 194; mechanical systems, 68–69; peacetime implementation, 23–24; policy implications, 4, 25–26; potential significance, 3–4, 7, 24, 31; prospects for nuclear disarmament, 143, 160–66; rationale, 7–8; recognition of and response to, 22–25; schools of thought, 11–12, 17–18; significance of, 17, 106–07; systems of systems school, 11, 12–13; technology, 2–3, 5–6, 11, 12–14, 20–21, 30, 104, 106, 107; vulnerability school, 16–17
Rich, Ben, 141
Robotics technology: in unmanned aerial vehicles, 76; microconstruction, 37–38, 43, 76; sensor technology and, 37, 43, 148
Rocket technology, 76–79
Roll-on/roll-off vessels, 178
Rosen, Stephen, 29
Rumsfeld, Donald, 96
Russia, 168
Rwanda, 150

Sailing ships, 20–21
Satellite technology, 27; antisatellite weapons, 55, 137; communications network, 52–53; global positioning system, 54–55; near-term needs, 176; prospects for technological development, 109–10; radar imaging, 38, 39–40, 48; reconnaissance, 120; use by U.S. adversaries, 61; visible light imaging, 34; vulnerability, 104
Saudi Arabia, 144, 145
Sealift, 149, 150, 177
Seismic monitoring, 42, 50–51

Sense and destroy armor munition, 81, 122

Sensor-fuzed weapon, 14, 87–88, 173, 174

Sensors, 27; acoustic, 44, 50, 67, 148–49; battlefield, 33, 42–43, 116–17; biological agent detection, 36, 43–44, 51, 164; chemical weapons detection, 43, 66; current capacity, 34–35; decoy detection, 35, 46; dominant battle-space knowledge, 11, 13–14; future battlefield needs, 148–49; limits of technological advance, 3, 45–51, 107, 148, 194; locating fired weaponry, 114; magnetic, 50, 67; prospects for technical development, 33, 66–67, 107, 109, 110–11, 132, 140–41, 194; radar and radio emission, 38–41, 46–50, 66–67; radioactive materials detection, 44, 51, 67, 162–63; range of applications, 34; RMA premise, 2, 106, 107; robotic transport, 37, 43, 148; seismic monitoring, 42, 50–51; underwater, 42, 44, 66; visible and near-visible light, 34–38, 45–46

Serbia, 2, 14, 28, 52–53, 74, 128–29, 146–47, 150, 155

Shalikashvili, John, 2, 19

Shelton, Henry, 155

Shinseki, Eric, 16

Snipers, 114–15, 148–49

Somalia peacekeeping action, 1–2, 27, 112; Mogadishu firefight, 118–19, 128

Sonar technology, 42, 44, 134, 135; prospects for development, 138–39

Soviet Union, 9

Space-based weapons, 16; antimissile lasers, 103; hypervelocity rods, 89

Spectroscopy, 43–44

Stark, U.S.S., 134

Stealth technologies, 27; antiradar missiles, 86–87; B-2 bomber, 146–47; for tanks, 83; historical development, 72; limitations, 47; prospects for technological development, 123; radar detection, 41

Stinger missile, 35

Strategic Defense Initiative, 103, 141

Strategic planning: anti-Clausewitzian, 8–9; current challenges, 4; DoD technology goals, 19–20; implications of RMA, 143; innovation without, 5; overseas basing, 144–53; prospects for technology-related change, 139–40; recognizing and responding to RMAs, 22–25; technology in, 1; use of arsenal ships, 147–48

Submarine-launched ballistic missiles, 89

Submarines, 134, 135; detection, 42, 44, 48–49, 138–39; propulsion systems, 79

Sun Tzu, 8–9

Surveillance technology. *See* Intelligence, surveillance and reconnaissance

Systems of systems concept, 11, 12–13, 173

Taiwan, 136, 196

Tanks, 29; antitank technology, 44, 81–82, 122, 127–28; current design characteristics, 81; current R&D efforts, 82–83; lightweight, 178; Persian Gulf War, 10; prospects for technological development, 123; radar countermeasures, 46–47; unmanned, 85

Terrorism, 179–80, 181

Theater high-altitude area defense (THAAD), 97–99, 100

Thermal imaging, 36

Toffler, Alvin, 17, 22

Toffler, Heidi, 17, 22

Tomahawk missile, 88

Transport technology, 17; aircraft, 73–74, 145, 178; blimps, 74, 151; current deployment capacity, 178; future forward-deployment patterns, 146, 149–51, 195–96; future needs, 177–78; ground vehicle design, 83–85, 113–14; historical RMAs, 20, 21; implications for infantry combat, 113, 149–51; mine-resistant vehicles, 178–79; prospects for technological development, 104, 139, 150–51; RMA premise, 2, 68–69, 104, 106; rocket propulsion, 77–79; sealift technology, 149, 150, 177; ships, 79–81, 134–35; vulnerabilities, 17, 113–14

Trident missiles, 76–77

Trumpet satellite, 39

Two-war strategy, 25, 173–74

U.S. military superiority: current status, 153, 168–69; deterrence value, 165–66; future of air power, 146; implications of RMA, 2, 3; relations with allies, 153–55; technological basis, 1; Ultraviolet sensors, 34

Underground targets, 90, 111; stores of fissile materials, 163

Unmanned aerial vehicles, 28, 34, 37, 57; advantages, 74; battlefield application, 116; for bomb targeting and delivery, 87, 88; research and development, 74–76; RMA shopping list, 158; role of, 74, 76

Unmanned ground vehicles, 85

Urban combat, 112, 117–18

V-22 Osprey, 73, 189–90

Van Riper, Paul K., 118

Vietnam, 1, 9, 27, 117

Vulnerability of U.S. forces: armored vehicles, 81–82; B-2 bomber, 146–47; communications systems, 59–61, 174–75, 194; computer technology, 62–63; cruise missile technology, 102; current status, 176–77; electronics systems, 124; fixed battlefield assets, 124; implications of RMA, 3, 4, 12, 16–17, 29–30, 107; in high-technology heavy combat, 125–28; missile technology of potential adversaries, 96; Mogadishu firefight, 118–19; naval, 79, 80–81, 110, 134, 135–39, 140; overseas bases, 151; technology-thwarting tactics, 117–18; to aircraft technology, 76; troop transport, 113–14

Warfare: future combat settings, 5; heavy combat, 119–28; infantry engagements in complex settings, 112; prospects for technological change, 108, 139; technology-thwarting tactics, 117–18

Weaponry: antiarmor, 81–82; computer simulations and modeling, 26, 63–64; countersniper technology, 114–15; electric guns, 83; explosives technology, 90; for heavy combat, 119–28; futuristic, 16; guided missiles, 85–87, 110; historical development of high-technology warfare, 9; historical RMAs, 20–21; kinetic energy, 89; near-term technology development goals, 173–74; near-term technology development investments, 172, 186–91; nonlethal, 91–93, 105, 111, 115, 119; precision, 120, 121; recent advances, 27–28; recommendations for research and development, 173–74; RMA premise, 3, 14, 106; RMA shopping list, 158; rocket delivery technology, 78; submunitions, 87–88; tank guns, 83; technological basis of RMAs, 6, 20; weapons of

mass destruction, 93–95, 165–66
Weapons performance: advances achieved
    without RMAs, 27–28; antiship, 135;
    antitank, 127–28; missile defense, 96,
    97–99; Persian Gulf War, 9–10;

prospects for technological develop-
    ment, 104–05, 109, 149
Welch, Larry, 100
World War II, 20, 23, 27, 29; blitzkrieg
    operations, 3, 8, 21